World Scientific Series on
Future Computing Paradigms
and Applications – **Vol. 4**

# Visual Quality Assessment and Enhancement for Natural Images

Surya Prakash

Indian Institute of Technology Indore, Indore, India

Piyush Joshi

Indian Institute of Information Technology, Sri City, Chittoor, India

**World Scientific**

NEW JERSEY · LONDON · SINGAPORE · BEIJING · SHANGHAI · HONG KONG · TAIPEI · CHENNAI · TOKYO

*Published by*

World Scientific Publishing Co. Pte. Ltd.

5 Toh Tuck Link, Singapore 596224

*USA office:* 27 Warren Street, Suite 401-402, Hackensack, NJ 07601

*UK office:* 57 Shelton Street, Covent Garden, London WC2H 9HE

Library of Congress Control Number: 2025001949

**British Library Cataloguing-in-Publication Data**
A catalogue record for this book is available from the British Library.

**World Scientific Series on Future Computing Paradigms and Applications — Vol. 4**
**VISUAL QUALITY ASSESSMENT AND ENHANCEMENT FOR NATURAL IMAGES**

Copyright © 2026 by World Scientific Publishing Co. Pte. Ltd.

ISBN 978-981-12-5728-5 (hardcover)
ISBN 978-981-12-5729-2 (ebook for institutions)
ISBN 978-981-12-5730-8 (ebook for individuals)

For any available supplementary material, please visit
https://www.worldscientific.com/worldscibooks/10.1142/12874#t=suppl

Desk Editors: Aanand Jayaraman/Amanda Yun

Typeset by Stallion Press
Email: enquiries@stallionpress.com

# About the Authors

**Dr. Surya Prakash** is currently working as a Professor in the Department of Computer Science and Engineering, Indian Institute of Technology Indore, India where he served as head of the department during 2017 to 2020. He has received his PhD, M.S., and B.Tech degrees, all in computer science and engineering, from Indian Institute of Technology Kanpur, India, Indian Institute of Technology Madras, India and University Institute of Engineering and Technology Kanpur, India respectively. His research interest includes image processing, computer vision, pattern recognition, machine learning, deep learning and biometric security. He has published several articles in refereed journals and conferences. He has also co-authored two books titled *Ear Biometrics: Localization and Recognition* published by Springer and *IT Infrastructure and Its Management* published by Tata McGraw Hills. He has become Sir Visvesvaraya Young Faculty Research Fellow (YFRF) in 2019. He is serving as a reviewer of several international journals and has been a member of the advisory boards, organizing committees and technical program committees of several international and national conferences and workshops. He has also organized several special sessions in international conferences and short-term courses in the domain of computer vision, machine learning and biometric security.

**Dr. Piyush Joshi** is currently working as an assistant professor at the Indian Institute of Information Technology (IIIT), Sri City, Chittoor. Prior to joining IIIT Sri city, he worked at the University of Birmingham, UK in Extreme Robotics Lab as a Faraday Institution Research Fellow. He has completed his PhD in the Discipline of Computer Science and Engineering from Indian Institute of Technology Indore, India. Before joining PhD, he received an M.Tech. degree in Information Technology (IT) from the Indian Institute of Information Technology, Allahabad (IIITA) and B.E. degree in Computer Science and Engineering from Rajiv Gandhi Proudyogiki Vishwavidyalaya (RGPV) Bhopal, India. His research interests include computer vision, image processing, artificial intelligence and pattern recognition. He has published several articles in refereed journals and conferences in these areas.

# Contents

# Chapter 1

# Introduction

There are numerous areas such as object recognition, medical applications, computational photography, industry automation, surveillance and access control, where digital images are being used and they play a significant role in representing useful information. However, the images may get distorted while passing through several operational steps such as image acquisition, transmission, compression and reconstruction in different operations and applications. For example, the presence of a very high or low contrast or poor illumination may severely affect the content of an image. Similarly, the use of a lossy compression technique while reducing the storage requirement of an image may insert blurring and ringing artifacts in the image. Further, while transmission of an image, some data may be lost because of the limited bandwidth of the transmission channel, and due to this, the quality of the image may degrade. The use of such distorted and poor-quality images may affect the performance of an image-based application. Therefore, identifying and quantifying degradation in images become very essential for controlling and enhancing a distorted image [1].

The quantification of the distortion present in an image can be carried out by estimating the quality of the image. There exist several ways of image quality assessment (IQA) to compute perceptual quality of an image in an automatic manner. These techniques offer ways to assess distortions of different kinds. In this book, we focus our discussion on a few critical image distortions such as the ones caused due to poor contrast (very low or very high contrast), poor illumination, noise and blur and discuss them in detail. Further, we also discuss ways to assess image quality in the presence of these distortions. In addition, the discussion covers quality-based image enhancement techniques for improving the content of images affected due to different distortions.

The subsequent part of the discussion in this chapter is organized as follows: Section 1.1 introduces different image quality attributes used for quality assessment. Section 1.2 presents a brief discussion on image quality assessment. Further, Section 1.3 describes image quality aware enhancement of images, whereas Section 1.5 presents various metrics used to evaluate image quality assessment and enhancement techniques.

## 1.1   Image Quality Attributes

Quality estimation of an image is carried out by analyzing various image attributes, such as contrast of the image, quality of illumination, presence of noise and noticeable blocking artifacts. These image quality attributes are briefly described in the following:

1. **Contrast:** Contrast is an important parameter which provides a good amount of information about the quality of an acquired image. The literal meaning of contrast is the difference. For an image, it is computed by estimating the difference between the darkest and the brightest spots present in the image. A large value for the difference shows high contrast, whereas a small value for the difference depicts low contrast. Both low as well as high contrasts are not good for image representation and are the cause of poor quality of an image. Figure 1.1 shows a few example images having low and high contrasts where in Figure 1.1(a), contrast is decreasing and in Figure 1.1(b), contrast increases when one moves from left to right.

2. **Illumination:** Illumination refers to the falling of light onto the surface of an object and making it visible as it is illuminated. For the acquisition of a good image through cameras, an adequate amount of illumination is required. The quality of an image may get degraded in the presence of undesirable illumination due to poor ambient lighting. For example, the presence of both very high illumination and very low illumination may degrade the quality of the captured image. It is observed that the presence of poor illumination causes dark or bright spots in the image. Figure 1.2 shows a few examples of images having poor quality due to the presence of bad illumination conditions.

(a) Decreasing contrast

(b) Increasing contrast

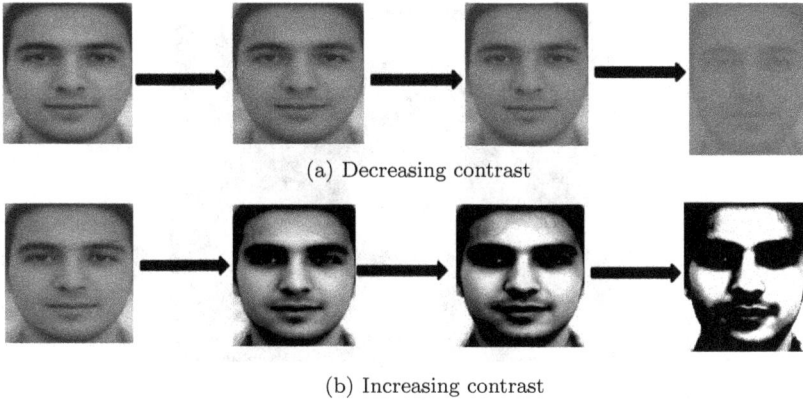

**Figure 1.1.** Examples of a few images with varying contrast.

3. **Noise:** The noise in an image refers to the random variation of brightness or color information in it. Primarily, the circuitry or sensor of the digital camera or scanner is responsible for the noise in the image. While capturing an image through a digital camera, the noise mainly shows up when shots are taken in low lighting conditions and sensors used in the camera are small. There are mainly two types of sensors, CCD (charge-coupled device) and CMOS (complementary metal oxide semiconductor), which are used in digital cameras. It is observed that CCD sensor-based cameras are more sensitive to light and produce better-quality images with low noise. On the other hand, CMOS sensor-based cameras relatively produce images with more noise and require more lighting to produce good images. However, the power requirement of CCD sensors is quite high as compared to CMOS sensors where a CCD sensor consumes roughly 100 times more power than a comparable CMOS sensor. Figure 1.3 shows a few examples of images having degradation due to the presence of noise.

4. **Artifacts in deblocked images:** Lossy data compression is often used in many applications to make the size of the digital images small so that they can be stored in a desired disk space or can be transmitted using the available bandwidth. JPEG (Joint Photographic Experts Group) is a commonly used lossy compression standard. At a low bit rate, it often produces blocking artifacts in images which is caused due to use of block-based encoding along with high quantization. The blocking artifact produces visible discontinuities in the compressed image and appears as unusually large pixel blocks. It is also sometimes

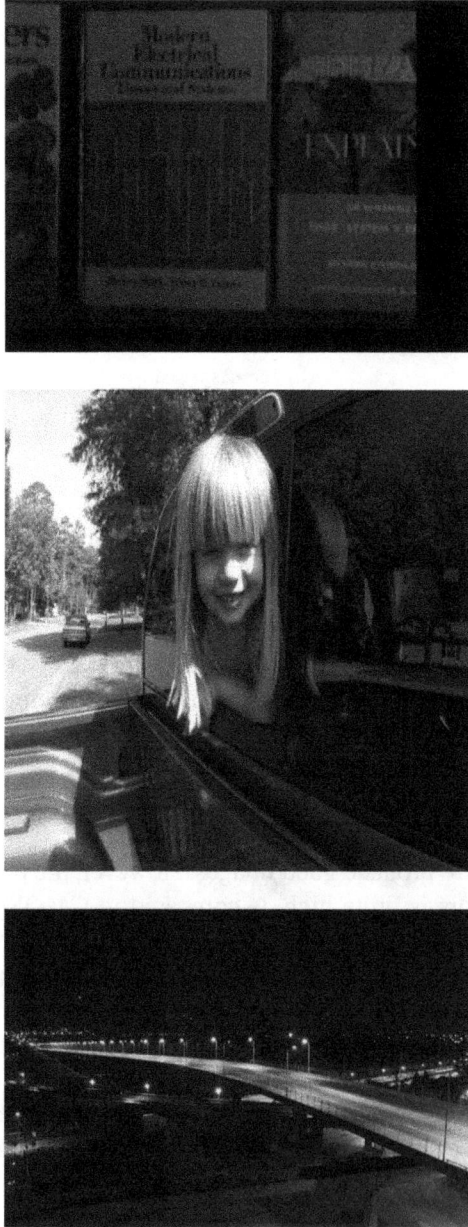

**Figure 1.2.**   Example of a few poor quality images due to bad illumination conditions [2].

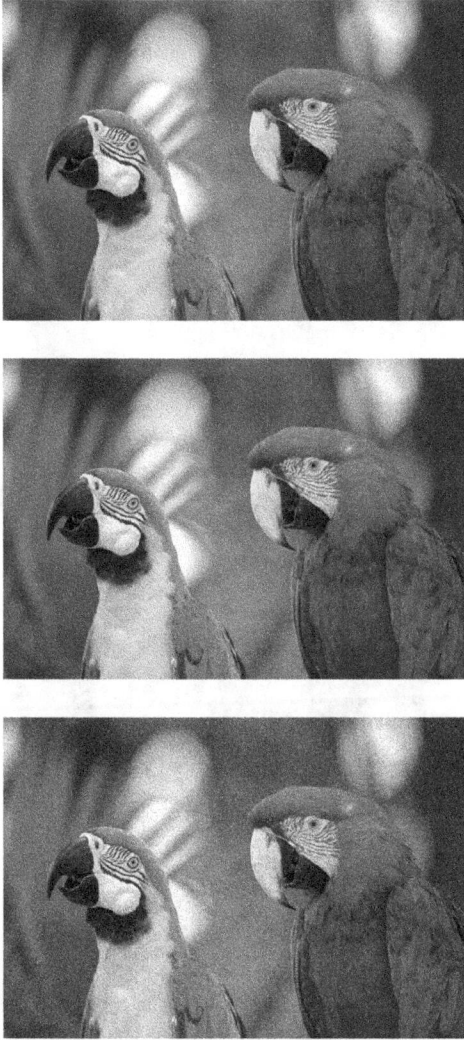

**Figure 1.3.** Example of a few noisy images where noise level in the images increases from top to bottom [3].

referred to as macroblocking. To remove the blocking artifacts in JPEG compressed images, deblocking is applied. However, in the process of deblocking, another distortion which is a blur is introduced, and as a result, two distortions get into a JPEG compressed image, that is, blocking artifacts and blur. Figure 1.4 shows a few examples of distorted images due to deblocking.

**Figure 1.4.** Examples of a few deblocked images where distortion level in the images increases from top to bottom [4].

5. **Sharpness:** Image sharpness is an important metric to measure the visual quality of an image. It is defined by the amount of information an image contains and gives us an idea about the clarity of textural detail present in the image. It is primarily constrained by camera configuration, magnification of the image and viewing distance. That means it is dependent on the sensor and lens parameters, such as design and manufacturing quality, focal length and aperture. In addition, sharpness is also affected by camera shake, atmospheric disturbances and focus accuracy.

## 1.2   Image Quality Assessment

An image-based application such as object recognition and surveillance require good-quality images to perform well as poor-quality images may degrade the efficiency of computer vision algorithms in these applications. If we assess the quality of the image before using it in the application, we can avoid the degradation in the performance of the algorithm by controlling or rejecting the poor-quality images. Subjective quality assessment is the most reliable way to evaluate quality of images. In this way of image quality assessment, an observer visualizes an image and provides a quality score for the image being assessed. More precisely, this method uses a large number of human observers called testers to assess the image quality. Generally, the testers are picked from different age groups, genders and occupations to get a robust and accurate assessment of the quality. Different testers provide scores for different images being analyzed. Subsequently, the obtained quality scores go through the data cleaning process, and finally, the average of the scores provided by different testers for an image is used as the final quality score of the image. Currently, the use of crowd-sourcing and crowd-workers [5] is heavily explored for obtaining the subjective quality assessment of images.

Although subjective quality assessment is indispensable, it poses certain challenges when it comes to its feasibility. In subjective quality assessment, the assessment procedure requires a lot of testers and the process is tedious as well as time-consuming. Further, the quality scores provided by the human observers are dependent on many factors such as viewing conditions, personal preferences, habits and others related to individual observers. To deal with these limitations of the subjective quality assessment procedure, objective quality assessment has come into existence in which the quality of an image is assessed in an automatic manner without using any human observer.

The goal of objective quality assessment is to get a model which is close to the human visual system in assessing the quality of the images. There are numerous techniques in the domain of objective quality assessment based on low-level vision [6–8], contrast [9–12], illumination [13], blur [14], noise [3], brain theory [15, 16] and statistics [17]. In the literature, the objective image quality assessment techniques are categorized into three types, namely no-reference (NR)-based quality assessment, reduced-reference (RR)-based quality assessment and full-reference (FR)-based quality assessment [18].

### 1.2.1    *FR-based image quality assessment*

In this type of method, an input distorted image is compared with the reference image that is considered to be distortion-free and to have perfect quality. There are a few widely used metrics for full reference-based quality assessment available in the literature. These are structural similarity index (SSIM), mean squared error (MSE) [19], visual information fidelity [20], multi-scale structural similarity index [21], feature similarity index [22] and most apparent distortion [23]. Chapter 6 elaborates more on FR-based image quality assessment techniques.

### 1.2.2    *RR-based image quality assessment*

In this type of quality assessment technique, only partial information about the reference image is required. Here, instead of using the complete reference image, extracted features from the reference image are utilized to assess the quality of an image. RR methods are further classified into three categories: The first category uses low-level statistical features of natural images for quality assessment. In the second category, sufficient information about the image distortion is considered available. The last category is based on the human visual system (HVS) where physiological and/or psychophysical studies are utilized. A detailed discussion on RR-based image quality assessment techniques is presented in Chapter 7.

### 1.2.3    *NR-based image quality assessment*

NR-based image quality assessment techniques aim to assess the quality of a given image without using any reference image. These techniques operate completely on the distorted image. In many real-world applications, a reference image is not available for the distorted image. This makes FR and RR-based image quality assessment techniques fail to estimate the quality of the image in such scenarios. This has drawn the attention of the researchers to work toward NR-based image quality assessment. Some widely used NR-based image quality assessment techniques can be found in BIQES [24], SA-SS [25], SA-ZC [25] and NOMDA [14], BLIINDS [26] and DIIVINE [27]. Chapter 7 discusses NR-based image quality assessment techniques in detail.

## 1.3   Quality Aware Image Enhancement

Image enhancement aims at improving the quality of images for a better display and for further use in an image-based application. The image enhancement techniques available in the literature can be categorized into two types: First, the spatial domain techniques which directly consider pixels in their computation, and second, the frequency domain techniques which work on the Fourier transform of the image. It is seen that most of the enhancement techniques, falling in either of these categories, directly apply their enhancement method on the image without knowing its content and perceptual quality and this leads to either over or under-enhancement of the image. This happens due to the fact that different images may be suffering from different types and amounts of distortions, and hence, blind application of an enhancement procedure on them may create image artifacts rather than enhancing them. For example, images which are acquired in a dark environmental settings may face low visibility and in turn, it may degrade the visual appearance of images and subsequently the performance of computer vision systems. On the other hand, images from medical domain may face very different challenges while acquiring and may have very unique issues to be overcome for improvement of their quality. These drawbacks of image enhancement techniques can be overcome if we make the enhancement process quality-aware. Let us say an image is being enhanced using a Wiener filter. To achieve good enhancement, an appropriate size should be chosen for the filter. A fixed-sized filter applied to images having different noise levels may not produce desirable results as less noisy images will require a filter of a smaller size, whereas highly noisy images will require a filter of a larger size. Hence, filter size needs to be adaptively chosen based on the amount of noise present in the image.

This can be achieved by first estimating the quality of an image and then computing the size of the Wiener filter based on the estimated quality of the image.

In quality-aware image enhancement, the first quality of the image which is to be enhanced is estimated, and based on the outcomes, appropriate action about enhancing the images is taken. For example, if the estimated quality is less than a pre-defined threshold, then the image is considered to be of poor quality and a suitable enhancement procedure is applied to it to enhance its quality. Otherwise, no action is taken on the image for its enhancement as it originally carries an acceptable quality. In the recent past, there have been a few works reported for quality-aware image enhancement. In Ref. [28], a quality-aware technique is being proposed for enhancing the contrast of an image. In another work presented in Ref. [10], automatic enhancement of the contrast is being carried out with the help of artificial bee colony [29] technique. In Ref. [11], a technique is proposed for automatic contrast enhancement where phase congruency and statistics information of image histogram are being utilized. Further, a quality-aware adaptive logarithmic enhancement technique is being proposed in Ref. [2] for the enhancement of natural images.

## 1.4   Evaluation Metrics for Subjective Image Quality Assessment

Subjective image quality assessment is a way to assess the visual quality of an image as judged by a human observer. It basically provides the perceptual assessment of a human viewer about the visual attributes of an image and is generally employed while preparing an image database for providing the ground truth quality value for each database image. The image quality ground truth thus provided is often used in the evaluation of objective image quality assessment techniques. As such, there is no metric to evaluate or judge the correctness of the subjective quality scores as these are the scores given by human observers based on their subjective visual assessment of the image. Assessing image quality will be easy if all human viewers perceive all visual attributes of an image equally. However, this is not the case as the process involves subjectivity due to the involvement of human viewers and their differences in perception. Hence, there are ways suggested to perform proper scoring of the image quality and normalize the quality scores to bring them to a common scale. For example, a human observer may choose to assign excellent, very good, good, fair, poor, etc. as quality grading to images. However, these grading results are shown untrustworthy in many researches [30] as the observers may use different quality scales for different images and even may vary for different distortion types. Further, this kind of evaluation may lead to unreliable assessment due to limited availability of rating options. Hence, the development of a proper scoring method becomes very essential in subjective quality assessment. To address this issue of variations of score in subjective assessment of quality of images, two techniques, namely Difference Mean Opinion Score (DMOS) and Z-score, are popularly used. These techniques are elaborated in the following.

### 1.4.1 *Difference mean opinion score (DMOS)*

In place of directly assessing the quality of each individual image independently, modern image quality assessment techniques suggest the use of a reference image for each test image (image whose quality is to be assessed) for quality evaluations and use the difference in the raw quality scores of reference and test images as the quality score of the test image. Formally, it is expressed as *difference mean opinion score (DMOS)* which is defined as the difference between the raw quality scores of the reference and test images. Mathematically, it is given as follows:

$$d_{i,j} = r_{i,ref}(j) - r_{i,j} \tag{1.1}$$

where $r_{i,j}$ is the raw score for the $j$-th image assessed by $i$-th subject (human viewer). Further, $r_{i,ref}(j)$ denotes the raw score given by the $i$-th subject to the reference image corresponding to the $j$-th test image.

### 1.4.2 *Z-score*

It is seen that different human viewers use different ways of rating (by adjectives) the images. To unify scales across observers, a linear transformation that uses mean and standard deviation is utilized. The outcome of this linear transformation is called Z-score, and it is mathematically defined as follows:

$$z_{i,j} = \frac{d_{i,j} - \overline{d_i}}{\sigma_i} \tag{1.2}$$

where scores used in this evaluation are all DMOS. Further, the mean score, $\overline{d_i}$, and the standard deviation of the scores, $\sigma_i$, are computed across all images that are rated by the $i$-th subject.

To provide reliable subjective quality assessment results, there are some international standards also proposed in the literature. These help in setting up the proper environment for evaluation. For example, these standards dictate what should be the viewing conditions, the contrast and brightness of the display devices used in experimentation, how to perform subjective experiments, etc. These are discussed in Chapter 5 in detail.

## 1.5  Evaluation Metrics for Objective Image Quality Assessment

Evaluation metrics play an important role in assessing the performance of an objective image quality assessment technique. An evaluation metrics used in analyzing the performance of an objective quality assessment use objective and subjective quality scores where objective quality score is estimated by the image quality assessment technique and subjective quality score is provided by a human observer. The subjective quality score is normally supplied by the image databases along with each image and is provided as a realigned difference mean opinion score (DMOS) or as mean opinion score (MOS) [31]. As discussed in the previous section, these scores are generated using manual observation of images by human observers. A number of evaluation metrics such as Spearman rank-order correlation coefficient (SROCC), root mean square error (RMSE) and Pearson correlation coefficient (CC) are used for the analysis of objective image quality assessment techniques. These metrics in a sense measure how well the objective quality scores generated by the image quality assessment technique match with the corresponding subjective quality scores.

According to the deliberations in the elaborate research work of the video quality experts group [31], there is a nonlinearity present in the subjective scores. Hence, it is essential to perform a nonlinear mapping of the objective quality scores and bring them on the same scale before using them in the computation of evaluation metrics. The work in Ref. [31] suggested the use of the following five-parameter-based logistic function to achieve the nonlinear mapping of the objective quality scores:

$$Q(x) = \beta_1 \, logistic(\beta_2, (x - \beta_3)) + \beta_4 x + \beta_5 \qquad (1.3a)$$

$$logistic(\tau, x) = \frac{1}{2} - \frac{1}{1 + exp(\tau x)} \qquad (1.3b)$$

where $x$ is the objective quality score and $Q(x)$ is the quality score obtained after performing the nonlinear mapping. Further, $\beta_k (k = 1, 2, 3, 4, 5)$ are the constants used in the logistic function and are computed by minimizing the sum of squared differences between the subjective and the mapped quality scores.

The following discussion elaborates on popular evaluation metrics, namely root mean square error (RMSE), mean absolute error (MAE), Pearson correlation coefficient (CC) and Spearman rank-order correlation coefficient (SROCC).

## 1.5.1 *Root Mean Square Error (RMSE)*

It is the standard deviation of the prediction errors where a prediction error measures how far an objective quality score is from its corresponding subjective quality score. It tells us how accurate the predicted quality (objective quality) of an image is and how much it deviates from the corresponding subjective quality score. It essentially shows how spread out are these prediction errors and measures the prediction accuracy of the image quality assessment technique. The lower value of RMSE indicates better performance of the image quality assessment technique. Mathematically, it is defined as follows:

$$RMSE = \left( \frac{1}{M_d} \sum_{i=1}^{M_d} (s_i - p_i)^2 \right)^{\frac{1}{2}} \tag{1.4}$$

where $s_i$ and $p_i$ are respectively the subjective image quality score and the mapped objective image quality score obtained after performing a nonlinear mapping on the objective quality score, for the $i$-th image in a dataset of size $M_d$.

## 1.5.2 *Mean Absolute Error (MAE)*

This measure is similar to RMSE, and likewise, it is also a measure of prediction accuracy. It tells us how accurate the predicted quality of an image is and how much it deviates from the corresponding subjective quality score. It is computed using the following equation:

$$MAE = \frac{1}{M_d} \sum_{i=1}^{M_d} |s_i - p_i| \tag{1.5}$$

where $s_i$ and $p_i$ are respectively the subjective image quality score and the mapped objective image quality score obtained after performing a nonlinear mapping on the objective quality score, for the $i$-th image in a dataset of size $M_d$.

The root mean square error and the mean absolute error are the two metrics which are largely used interchangeably for evaluating the performance of objective image quality assessment techniques. However, root mean square error is often found less robust as compared to mean absolute error, as squaring of the difference values amplifies the score differences in root mean square error, whereas in mean absolute error, equal importance is given to all the differences.

### 1.5.3   *Pearson Correlation Coefficient (CC)*

It is the linear correlation coefficient between the objective and the subjective quality scores. It shows the prediction accuracy of the image quality assessment technique where the higher value of the coefficient indicates better performance. The computation of CC also uses the non-linear mapping of the objective quality scores before using them in the computation. The CC is formally given as follows:

$$CC = \frac{\sum_{i=1}^{M_d}(p_i - \overline{p})(s_i - \overline{s})}{\sum_{i=1}^{M_d}((p_i - \overline{p})^2)^{\frac{1}{2}}\sum_{i=1}^{M_d}((s_i - \overline{s})^2)^{\frac{1}{2}}} \tag{1.6}$$

where $s_i$ and $p_i$ are respectively the subjective image quality score and the mapped objective image quality score obtained after performing a nonlinear mapping on the objective quality score, for the $i$-th image in a dataset of size $M_d$.

### 1.5.4   *Spearman rank-order correlation coefficient (SROCC)*

It is a metric which is commonly used to understand the prediction monotonicity. A monotonic relationship between two related variables exists when as the value of one variable increases, the value of the other variable also increases or remains constant and vice versa. A few examples of monotonic and non-monotonic relationships are shown in Figure 1.5. The prediction monotonicity dictates the degree to which the quality predictions of a model (objective quality scores) correlate with the relative

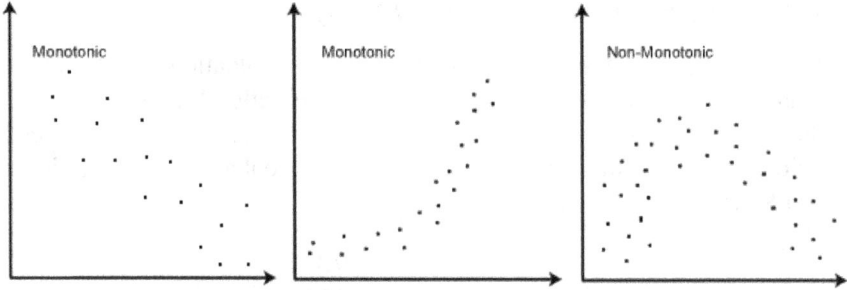

**Figure 1.5.** A few examples of monotonic and non-monotonic relationships.

magnitudes of the subjective quality scores. A higher value of SROCC indicates better performance of an image quality assessment technique. Mathematically, the SROCC is defined as follows:

$$SROCC = 1 - \frac{6\sum_{i=1}^{M_d} d_i^2}{M_d(M_d^2 - 1)} \qquad (1.7)$$

where $d_i$ indicates the difference between ranks of $i$-th image in the subjective and the objective quality assessment experiments. $M_d$ represents the size of the considered image database. The SROCC score lets us know the extent to which the image quality evaluation technique is able to capture the order of distorted images in accordance with the subjective scale. That means it examines the technique to find out the limit to which the objective quality scores generated by it agree with the relative magnitude of the subjective quality scores. It is worth noting that SROCC is independent of any nonlinear mapping that may exist between the subjective and objective quality scores.

An efficient image quality assessment technique needs to have higher values for CC and SROCC whereas low values for RMSE and MAE.

## 1.6   Evaluation Metrics for Image Enhancement

This section discusses two important metrics, namely peak signal-to-noise ratio (PSNR) and structural similarity (SM), which are used for the evaluation of image quality enhancement techniques.

### 1.6.1   *Peak signal-to-noise ratio (PSNR)*

One of the popular metrics that is used for evaluation of an image enhancement technique is peak signal-to-noise ratio (PSNR) [19]. The value of PSNR is computed between a reference image and its enhanced version to evaluate the performance of the image enhancement technique. The PSNR is formally defined as follows:

$$PSNR = 10 \cdot \log_{10} \left( \frac{MAX_I^2}{MSE} \right) \tag{1.8}$$

where $MAX_I$ is the maximum possible pixel value in the image $I$ and $MSE$ indicates the mean squared error. The higher value of PSNR shows better performance of the enhancement technique. It is to be noted that the computation of PSNR requires a reference image.

### 1.6.2   *Structural similarity index measure (SSIM)*

It is another widely used metric to assess the performance of an image enhancement technique. It is a full reference measure that computes the similarity between two images (the original distorted image and its enhanced version) by estimating their perceived qualities. SSIM estimates degradation in the image quality by measuring the change in structural information. It is computed by considering various windows in the image and is defined as follows:

$$SSIM(x,y) = \frac{(2\mu_x\mu_y + C_1) + (2\sigma_{xy} + C_2)}{(\mu_x^2 + \mu_y^2 + C_1)(\sigma_x^2 + \sigma_y^2 + C_2)} \tag{1.9}$$

where $x$ and $y$ are two windows of size $N \times N$. Further, $\mu_x$ and $\mu_y$ define the average of windows $x$ and $y$, respectively, whereas $\mu_x^2$ and $\mu_y^2$ are the corresponding variances of these windows. Next, $\sigma_{xy}$ defines the value of covariance for variables $x$ and $y$, and $C_1$ and $C_2$ are the two variables used to stabilize the division with a weak denominator.

### 1.6.3   *Lightness order error (LOE)*

It measures the naturalness in an image and is mostly used to analyze the performance of an illumination enhancement technique [32]. It considers an input image and corresponding enhanced output image for computation. Let $I$ be the original input image and $I_e$ be the corresponding enhanced output image. The lightness $L(x,y)$ of the image $I$ at pixel $(x,y)$ is defined as the maximum of its three color channels at that pixel as follows:

$$L(x,y) = \max_{c \in \{r,g,b\}} I^c(x,y) \tag{1.10}$$

Further, between pixel $(x,y)$ of the original image $I$ and its corresponding pixel in the enhanced image $I_e$, the relative order difference of the lightness is given as follows:

$$RD(x,y) = \sum_{i=1}^{m} \sum_{j=1}^{n} [U(L(x,y),L(i,j)) \oplus U(L_e(x,y),L_e(i,j))] \tag{1.11}$$

where

$$U(x,y) = \begin{cases} 1, & \text{if } x > y \\ 0, & \text{otherwise} \end{cases} \tag{1.12}$$

where $m$ and $n$ are the height and the width of image $I$, $U(x,y)$ is the unit step function and $\oplus$ is the exclusive operator. Based on the above, the LOE metric can be defined as follows:

$$LOE = \frac{1}{m * n} \sum_{i=1}^{m} \sum_{j=1}^{n} RD(i,j) \tag{1.13}$$

From Equation 1.13, it is clear that the smaller value of LOE indicates that the enhanced image consists of a naturally pleasing appearance and preserves a good amount of lightness order.

## 1.7 Contributions of this Book

This book aims to provide a comprehensive discussion on image quality assessment and enhancement along with the techniques proposed for these purposes. It presents recent advances in the domain of image quality assessment and quality-aware enhancement of distorted images. In this book, presented quality assessment and enhancement techniques are mainly focused on image distortions due to poor contrast, poor illumination, noise and blur.

This book consists of a total of 12 chapters. A brief description of the content of each chapter is as follows. Chapter 2 presents an introduction to the human visual system which lays the foundation for the subsequent discussion. It is important to understand the functioning and characteristics of the human visual system for a better understanding of image quality assessment and enhancement techniques. Chapter 3 explains the meaning of image quality and the degradation in the image quality. Chapter 4 presents a review of image quality assessment databases. It contains information on single and multiple distortion databases for IQA. Chapter 5 discusses the subjective and the objective image quality assessment. Chapter 6 discusses various full reference-based image quality assessment techniques; whereas, in Chapter 7, reduced reference-based image quality assessment techniques are presented. In Chapter 8, image quality assessment techniques that do not use any reference image are presented. Chapter 9 introduces quality-aware enhancement in images and discusses various techniques existing in the literature for this purpose. Chapter 10 presents the applications of image quality assessment. Chapter 11 contains information regarding some significant challenges in front of IQA techniques. In Chapter 12, emerging trends and future directions on image quality assessment and quality-aware image enhancement are discussed and the book is concluded.

# References

[1]  W. Lin and C.-C. J. Kuo, "Perceptual visual quality metrics: A survey," *Journal of Visual Communication and Image Representation*, 22(4), pp. 297–312, 2011.

[2]  P. Joshi and S. Prakash, "Image enhancement with naturalness preservation," *Visual Computer*, 36(1), 71–83, 2020.

[3]  P. Joshi and S. Prakash, "Nr-iqa for noise-affected images using singular value decomposition," *IET Signal Processing*, 13(2), 183–191, 2019.

[4]  P. Joshi, S. Prakash, and S. Rawat, "Continuous wavelet transform-based no-reference quality assessment of deblocked images," *The Visual Computer*, 34, 1739–1748, 2017.

[5]  D. Saupe, F. Hahn, V. Hosu, I. Zingman, M. Rana, and S. Li, "Crowd workers proven useful: A comparative study of subjective video quality assessment," in *Proceedings of the QoMEX 2016 8th International Conference on Quality of Multimedia Experience, Lisbon, Portugal*, 2016.

[6]  L. Zhang, L. Zhang, X. Mou, and D. Zhang, "FSIM: A feature similarity index for image quality assessment," *IEEE Transactions on Image Processing*, 20, 2378–2386, 2011.

[7] A. Liu, W. Lin, and M. Narwaria, "Image quality assessment based on gradient similarity," *IEEE Transactions on Image Processing*, 21, 1500–1512, 2012.

[8] G. Yue, C. Hou, K. Gu, S. Mao, and W. Zhang, "Biologically inspired blind quality assessment of tone-mapped images," *IEEE Transactions on Industrial Electronics*, 65, 2525–2536, 2018.

[9] S. Wang, K. Ma, H. Yeganeh, Z. Wang, and W. Lin, "A patch-structure representation method for quality assessment of contrast changed images," *IEEE Signal Processing Letters*, 22(12), 2387–2390, 2015.

[10] P. Joshi and S. Prakash, "An efficient technique for image contrast enhancement using artificial bee colony," in *IEEE International Conference on Identity, Security and Behavior Analysis (ISBA 2015)*, 2015, pp. 1–6.

[11] K. Gu, S. Wang, G. Zhai, W. Lin, X. Yang, and W. Zhang, "Analysis of distortion distribution for pooling in image quality prediction," *IEEE Transactions on Broadcasting*, 62(2), 446–456, 2016.

[12] K. Gu, L. Li, H. Lu, X. Min, and W. Lin, "A fast reliable image quality predictor by fusing micro- and macro-structures," *IEEE Transactions on Industrial Electronics*, 64(5), 3903–3912, 2017.

[13] S. Wang, J. Zheng, H.-M. Hu, and B. Li, "Naturalness preserved enhancement algorithm for non-uniform illumination images," *IEEE Transactions on Image Processing*, 22(9), 3538–3548, 2013.

[14] P. Joshi and S. Prakash, "Retina inspired no-reference image quality assessment for blur and noise," *Multimedia Tools and Applications*, 76(18), 18 871–18 890, 2017.

[15] K. Gu, G. Zhai, X. Yang, and W. Zhang, "Using free energy principle for blind image quality assessment," *IEEE Transactions on Multimedia*, 17, 50–63, 2015.

[16] G. Zhai, X. Wu, X. Yang, W. Lin, and W. Zhang, "A psychovisual quality metric in free-energy principle," *IEEE Transactions on Image Processing*, 21, 41–52, 2012.

[17] K. Gu, J. Zhou, J.-F. Qiao, G. Zhai, W. Lin, and A. C. Bovik, "No-reference quality assessment of screen content pictures," *IEEE Transactions on Image Processing*, 26, 4005–4018, 2017.

[18] H. R. Wu and K. R. Rao, *Digital Video Image Quality and Perceptual Coding*. CRC press, 2005.

[19] Z. Wang, A. Bovik, H. Sheikh, and E. Simoncelli, "Image quality assessment: from error visibility to structural similarity," *IEEE Transactions on Image Processing*, 13(4), 600–612, 2004.

[20] H. Sheikh and A. Bovik, "Image information and visual quality," *IEEE Transactions on Image Processing*, 15(2), 430–444, 2006.

[21]   Z. Wang, E. Simoncelli, and A. Bovik, "Multiscale structural similarity for image quality assessment," in *Proceedings of the Thrity-Seventh Asilomar Conference on Signals, Systems and Computers*, 2, 1398–1402, 2003.

[22]   L. Zhang, L. Zhang, X. Mou, and D. Zhang, "FSIM: A feature similarity index for image quality assessment," *IEEE Transactions on Image Processing*, 20(8), 2378–2386, 2011.

[23]   E. C. Larson and D. M. Chandler, "Most apparent distortion: full-reference image quality assessment and the role of strategy," *Journal of Electronic Imaging*, 19(1), 011 006(1–21), 2010.

[24]   A. Saha and Q. Wu, "Utilizing image scales towards totally training free blind image quality assessment," *IEEE Transactions on Image Processing*, 24(6), 1879–1892, June 2015.

[25]   J. Zhang, T. M. Le, S. Ong, and T. Q. Nguyen, "No-reference image quality assessment using structural activity," *Signal Processing*, 91(11), 2575–2588, 2011.

[26]   M. Saad, A. Bovik, and C. Charrier, "Blind image quality assessment: a natural scene statistics approach in the DCT domain," *IEEE Transactions on Image Processing*, 21(8), 3339–3352, 2012.

[27]   A. Moorthy and A. Bovik, "Blind image quality assessment: From natural scene statistics to perceptual quality," *IEEE Transactions on Image Processing*, 20(12), 3350–3364, 2011.

[28]   B. Subramanyam, P. Joshi, M. K. Meena, and S. Prakash, "Quality based classification of images for illumination invariant face recognition," in *Proceedings of IEEE International Conference on Identity, Security and Behavior Analysis (ISBA)*, 2016, pp. 1–6.

[29]   D. Karaboga and B. Basturk, "Artificial bee colony (abc) optimization algorithm for solving constrained optimization problems," in *Proceedings of International Fuzzy Systems Association World Congress (IFSA 2007), Foundations of Fuzzy Logic and Soft Computing, LNCS vol. 4529*, 2007, pp. 789–798.

[30]   A. M. van Dijk, J.-B. Martens, and A. B. Watson, "Quality assessment of coded images using numerical category scaling," in *Advanced Image and Video Communications and Storage Technologies*, vol. 2451, International Society for Optics and Photonics. SPIE, 1995, pp. 90–101.

[31]   "VQEG . (2009). Final report from the video quality experts group on the validation of reduced-reference and no-reference objective models for standard definition television, phase-I. Available: /http://www.vqeg.org/s."

[32]   S. Wang, J. Zheng, H. M. Hu, and B. Li, "Naturalness preserved enhancement algorithm for non-uniform illumination images," *IEEE Transactions on Image Processing*, 22(9), pp. 3538–3548, 2013.

# Chapter 2

# The Human Visual System

Human Visual System (HVS) [1,2] is an important part of the human body and is a crucial tool to know and understand the real world. Over the last few decades, there has been tremendous growth in understanding the HVS and researchers now knowing considerable insights about different parts of HVS. Researchers have extensively studied specific regions of the human visual system (HVS), particularly the cortical areas located in the cerebral cortex, to gain deeper insights into its functionality. For example, primary visual cortex $(V_1)$ which is the primary cortical region of the brain and receives most visual information from the retina has been studied by several researchers in detail in the past. It has been established in these researches that cortical areas of the brain are the reason for the high-level processing of visual data. Other than primary visual cortex $(V_1)$, there are several other cortical areas which have been identified in the brain. These areas are believed to be further involved in the higher-level processing of sensory data.

The study of the human visual system has advanced remarkably over centuries, evolving from early philosophical theories to intricate scientific explorations. This journey of discovery has deepened our understanding of the mechanics, physiology and neurological pathways involved in human vision. Here, we present the important historical progress of this fascinating field:

1. **Ancient Foundations and Philosophical Theories (5th Century BCE–2nd Century CE)**: In ancient Greece, philosophers such as Plato (427–347 BCE) and Aristotle (384–322 BCE) laid the earliest

foundations of visual theory. Plato speculated that the eyes emit rays that interact with the external world, producing vision. Aristotle countered this by suggesting that vision is dependent on light reflecting off objects into the eye, a more accurate concept that paved the way for future inquiry.

The Roman physician Galen (129–216 CE) made some of the earliest attempts to understand the anatomy of the eye, describing its major components, such as the lens, optic nerve and retina. Galen suggested that vision was linked to the brain — a critical shift toward understanding visual perception as a neuro-physiological process.

2. **Golden Age of Optics**: Ibn al-Haytham (965–1040 CE), widely regarded as the father of optics, produced *Kitab al-Manazir* (*The Book of Optics*), challenging ancient Greek theories. He posited that light enters the eye rather than emanates from it, and he conducted experimental research on the behavior of light, reflection and refraction, laying the groundwork for modern optics.

3. **Renaissance Contributions and Advances in Anatomy (14th–17th Century CE)**: With the Renaissance, scientific inquiry surged. Artists such as Leonardo da Vinci (1452–1519) dissected human eyes and documented their observations, advancing the anatomical knowledge of the eye. Andreas Vesalius (1514–1564) later made comprehensive studies of human anatomy, further refining knowledge about the eye's structure.

Kepler, a German astronomer, discovered that images are inverted as they pass through the lens, with the retina serving as the screen where images form. He compared the eye to a camera obscura, an early conceptualization that the eye functions similarly to a camera in forming images.

4. **Rise of Psychophysics and Visual Physiology (19th Century)**: The 19th century saw the emergence of psychophysics, a discipline studying the relationship between physical stimuli and sensory perception. Hermann von Helmholtz (1821–1894) conducted extensive research on the mechanics of vision, including the nature of color perception. He proposed the trichromatic theory of vision, which posits that the retina contains three types of receptors sensitive to red, green and blue light.

Concurrently, Ewald Hering proposed the opponent-process theory, which suggested that color vision operates through pairs of opposing colors: red-green, blue-yellow and black-white. The interaction between Helmholtz and Hering's theories laid the foundation for modern color vision science.

In the late 19th century, scientists identified rods and cones within the retina. Rods are highly sensitive to low light, while cones are responsible for color detection, explaining fundamental differences in day and night vision.

5. **Advances in Neurophysiology and Understanding Visual Pathways (Early 20th Century)**: By the early 20th century, neuroanatomists began to unravel how visual information travels from the eye to the brain via the optic nerve and is processed in the visual cortex. Research showed that different parts of the brain handle various visual functions, such as motion, color and depth perception.

   Gestalt psychologists, such as Max Wertheimer, introduced the concept that humans perceive objects as unified wholes rather than as individual parts. This shift influenced later understanding of how the brain organizes visual information, a principle still relevant in visual perception studies.

6. **Breakthroughs in Visual Cortex Functionality (Mid-20th Century)**: In the 1950s and 1960s, David Hubel and Torsten Wiesel conducted groundbreaking experiments on the visual cortex, discovering that specific neurons respond to particular features, such as edges and movement. They proposed a hierarchical model of visual processing in which the brain builds a complete visual representation from simpler elements — a concept foundational to understanding the brain's visual architecture. Their work earned them the Nobel Prize in 1981.

   The development of retinotopic mapping techniques helped map how visual space is organized in the brain. This concept, which involves each part of the retina corresponding to a specific area in the visual cortex, was key to understanding the visual cortex's organization.

7. **Neuroimaging and Computational Vision (Late 20th and 21st Centuries)**: The advent of functional magnetic resonance imaging (fMRI) allowed researchers to observe real-time visual processing in the human brain, significantly advancing our understanding of brain areas involved in visual perception.

   Inspired by human vision, researchers developed artificial neural networks that simulate the brain's visual processing pathways, particularly through convolutional neural networks (CNNs). These models have advanced fields such as object recognition and have applications in both neuroscience and machine vision.

   Studies in neuroplasticity have revealed the brain's adaptability in cases of sensory loss or impairment, showing how individuals adapt to visual changes. This has led to explorations in visual prosthetics, such as bionic eyes, and retinal implants for restoring vision in those with severe impairments.

Visual physiology [3] and psychophysics [4] are the two areas where significant research findings have been obtained to understand the HVS. Despite advancements in understanding the HVS, there still exist a few challenges. For example, there are no exact operational explanations of cortical areas, and therefore, this is currently the most explored area of research.

Research in the area of visual physiology has shown the existence of four important pillars of visual processing in the brain. These include retina processing [5], optical processing [6], lateral geniculate nucleus (LGN) processing [7] and visual cortex processing [8].

Figure 2.1 shows the basic steps involved in processing in the human visual system. The function of the eyes is to complete optical processing and this is similar to a camera. The sclera in eyes is equal to a spherical camera bellows, iris is like an aperture used to regulate brightness in the retina, lens is similar to lens in a camera and the retina acts like film in the camera. Retina is responsible for forming an object image when light focuses on it [5]. Retina consists of three categories of cells named bipolar cells, photoreceptor cells and ganglion cells. They are responsible for photoelectric conversion and transmission of information. Optical signal received from eyeballs is converted into an electrical signal. This signal is now passed to LGN which controls the amount of information by applying a threshold. In the end, the visual cortex is used to do many higher-level tasks, such as identifying objects, understanding, depth perception, and visualising color and motion.

The advancement in physiological research [7] motivates the development of visual psychophysics. Visual psychophysics is the study of the relationship among visual psychological phenomena using mathematical models and measurement techniques. Many visual psychophysics phenomena include hypothetical models due to the complexity of HVS. A few models have been developed that show the low-level processing of visual psychological characteristics, including contrast sensitivity [9], visual attention [10], masking effects [11] and luminance nonlinearity [12]. Luminance nonlinearity explains information about a human judgment of the brightness of an object. It states that human has a poor understating of absolute brightness (low contrast sensitivity) and strong if relative

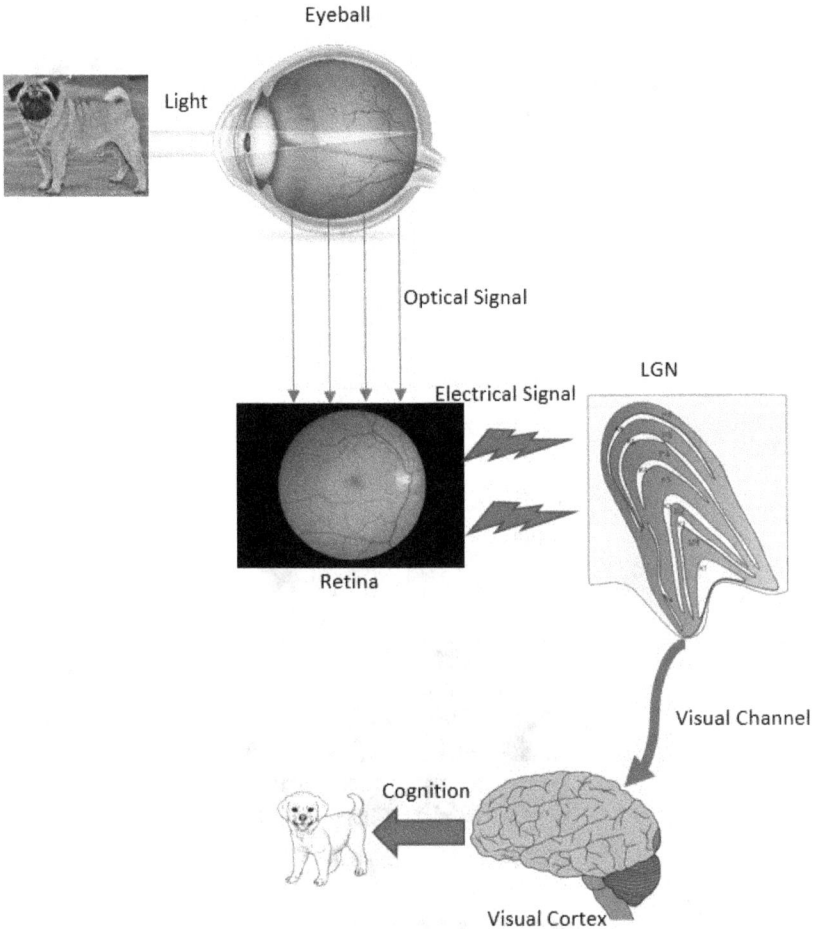

**Figure 2.1.** The process occurred in HVS.

differences in brightness (high contrast sensitivity) are present. Contrast sensitivity describes frequency response characteristics of the HVS. It is the ability of the human eyes to distinguish the differences in intensity. Contrast sensitive function (CSF) is given by Campbell and Robson [13] which explicates the presence of band-pass filter in HVS. Masking effect [14, 15] explains the effect on one component of an image (stimulus A) due to the presence of another component (stimulus B). Some advanced studies show that HVS is a multi-channel parallel [16] concept where different neural channels process different visual information. Further, the different

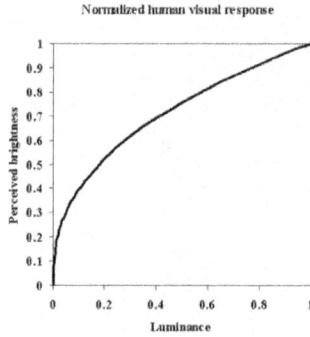

(a) Luminance nonlinearity

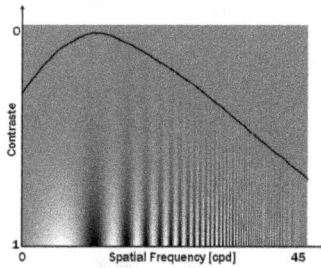

(b) Contrast sensitivity

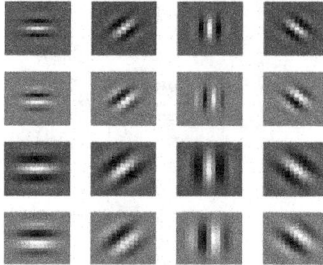

(c) Multiple channel

(d) Visual attention

**Figure 2.2.** Representation of some visual psychophysics concepts [23].

processed information is analyzed by different cortical cells using different spatial frequencies. For example, most of the neurons in the primary visual cortex are sensitive to visual stimuli with specific spectrum locations, frequency and orientation. Visual attention is the cognitive process that extracts only important information from the observed scene. This study shows that certain interesting spots act as the representation of an image. There are two important strategies developed by James [17] named bottom-up saliency and top-down saliency. Bottom-up saliency approach (Itti [18], GBVS [19], Gaffe [20]) is based on the stimulation of the low-level features of the target. On the other hand, top-down saliency approach (Searchlight [21] and Tsotsos [22]) depends on the visual processing tasks. James [17] presented a bottom-up saliency strategy in the field of Principles of Psychology. Techniques involved in this concept are Itti, GBVS and Gaffe. These techniques are based on feature integration that focuses on stimulation of the low-level features of the target. Also, they emphasized on attention process that impacts target features and affects the observer's knowledge and experience. Figure 2.2 presents a few visual psychophysics phenomena.

## 2.1 Psychophysical HVS Features

In this section, we present some psychophysical HVS features used in the development of image quality assessment techniques.

### 2.1.1 *Foveal and peripheral vision*

The densities of cone cells and ganglion cells in the retina vary, peaking at the fovea and decreasing rapidly with distance from it. Consequently, when a human observer fixates on a point in their environment, the region around the fixation point is perceived to have the highest spatial resolution, while resolution decreases with distance from the fixation point. This phenomenon, known as foveal vision, results in high-resolution vision at fixation points and progressively lower resolution in peripheral vision. While most image quality assessment models focus on foveal vision, some also incorporate peripheral vision. Additionally, models may resample images to match the receptor density in the fovea, aiming to better approximate the human visual system and enhance model calibration.

## 2.1.2   *Light adaptation*

The human visual system functions effectively across a broad spectrum of light intensity, encompassing diverse conditions from dim moonlit nights to bright sunny days. This capability is facilitated by a phenomenon termed light adaptation, which regulates the amount of light entering the eye via the pupil and adjusts the gain of post-receptor neurons in the retina through adaptation mechanisms. As a result, the retina encodes the contrast of visual stimuli rather than absolute light intensities. Weber's law maintains the contrast sensitivity of the HVS across varying background light intensities.

## 2.1.3   *Contrast sensitivity functions*

The Contrast Sensitivity Function (CSF) characterizes the human visual system's sensitivity to various spatial and temporal frequencies within visual stimuli. This sensitivity variation may stem from the receptive field properties of ganglion cells and cells in the LGN, or from internal noise characteristics of HVS neurons. Consequently, some HVS models incorporate CSF as a filtering process, while others utilize weighting factors for sub-bands following frequency decomposition. Although CSF varies with distance from the fovea, for foveal vision, spatial CSF is typically represented as a space-invariant band-pass function. While CSF tends to be slightly band-pass in nature, most quality assessment algorithms adopt a low-pass version, enhancing robustness to changes in viewing distance.

## 2.1.4   *Masking and facilitation*

Masking and facilitation play crucial roles in the Human Visual System (HVS) by modeling interactions between different image components located at the same spatial position. Masking or facilitation refers to the phenomenon where the presence of one image component (the mask) decreases or increases the visibility of another image component (the test signal). Typically, the mask reduces the visibility of the test signal compared to when the mask is absent, although in some cases, it may also facilitate detection. The masking effect is typically the strongest when the mask and the test signal have similar frequency content and orientations. Most quality assessment methods integrate either a masking or facilitation model, while some incorporate both.

### 2.1.5 *Pooling*

Pooling involves consolidating the outputs of visual streams to derive a single quality measurement or visibility decision regarding artifacts. The exact mechanism of pooling in the Human Visual System (HVS) remains unclear, although it likely involves cognition, as certain distortions may be more bothersome in specific areas of a scene, such as human faces. Despite this, most quality assessment metrics utilize Minkowski pooling to combine error signals from various frequency and orientation-selective streams, as well as across spatial coordinates, to determine fidelity.

## 2.2 Involvement of HVS in Image Quality Assessment

In this section, we explore image quality assessment techniques based on HVS. These techniques consider the most important properties of HVS in developing an HVS-based IQA. The properties of HVS that should be considered for building an image quality assessment model are as follows:

1. **Contrast Sensitivity**: The HVS is highly sensitive to contrast, especially at certain spatial frequencies, which means an IQA model should account for how contrast impacts perceived image quality. Higher contrast at edges and textures is often perceived as higher quality. Here are some examples related to contrast:

   (a) *Reading in Low-Contrast Texts*: When reading light gray texts on a white background, it can be challenging to discern the letters, especially if the text is small. This example highlights how lower contrast reduces visibility, making it harder for the HVS to distinguish details.

   (b) *Foggy or Low-Light Environments*: In a foggy scene, objects like road signs or cars appear with reduced contrast, and the HVS struggles to detect the details. People are more sensitive to medium-contrast objects (like dark on light or vice versa) than low-contrast ones, especially when viewing from a distance.

   (c) *Images with High vs. Low Contrast*: High-contrast images, such as a black-and-white line drawing, are easier to see and interpret because the contrast between light and dark areas enhances visibility. Conversely, in a low-contrast photo, such as a hazy sunset, details appear less sharp, showing the HVS's reduced sensitivity to fine distinctions at lower contrast levels.

(d) *Contrast Sensitivity in Image Processing*: In applications such as medical imaging, contrast is enhanced (e.g., for X-rays or MRIs) to make subtle details more visible to the eye, as the HVS is better at detecting details with heightened contrast.

2. **Spatial Frequency Sensitivity**: The HVS reacts differently to varying spatial frequencies, responding best to mid-range frequencies. IQA models should prioritize frequencies that humans are more likely to note, particularly for distortions and loss of detail. Here are examples that illustrate this:

(a) *Textured Fabrics and Patterns*: When looking at textured fabrics, such as stripes or checker patterns, the HVS can easily detect medium-sized patterns (around 4–8 cycles per degree of visual angle) but may struggle to see very fine, closely packed lines or very large, spaced-apart patterns. Medium-frequency textures are more visually prominent, whereas finer or larger patterns may appear blurred or indistinct.

(b) *Facial Recognition*: Humans are especially good at detecting facial features, which tend to contain medium spatial frequencies. Details such as the eyes, nose and mouth provide enough contrast and are spaced at a frequency that the HVS can detect readily, which aids in facial recognition. Conversely, very fine details (e.g., tiny skin pores) or large, generalized shapes are less critical to recognition because they fall outside the HVS's peak spatial frequency sensitivity.

(c) *Highway Signs*: Road signs are designed with bold, medium-sized letters that align with the HVS's peak spatial frequency sensitivity, making them easy to read from a distance. If the letters were too fine or too large and widely spaced, they would be harder to read due to the decreased sensitivity of the HVS at those frequencies.

(d) *Medical Imaging*: In radiology, images are adjusted for contrast at specific spatial frequencies to highlight structures of interest. For example, bone and soft tissue details in X-rays are often presented at frequencies that the HVS detects well, enhancing diagnostic accuracy by making the right details more visible.

3. **Color Sensitivity**: Humans perceive color based on how receptors in the eye process red, green and blue. The model should weigh colors differently, giving more importance to colors humans are more sensitive to, like green, and adjust for reduced sensitivity to color distortions in the periphery of vision. Here are examples that highlight this:

   (a) *Greater Sensitivity to Green*: The HVS is most sensitive to wavelengths around the green spectrum (around 555 nm), which corresponds to the middle of the visible spectrum. This is why many displays, lighting systems and safety signs use green as it appears bright and stands out easily. Additionally, humans can distinguish more shades of green than other colors, a trait believed to have evolved for detecting foliage and other natural elements.

   (b) *Red and Blue Perception Differences*: Humans are less sensitive to red and blue light wavelengths than to green. For instance, small red or blue details in a scene may appear less prominent or blurrier than green details of the same size. This reduced sensitivity is one reason red and blue often appear less sharp and vibrant, especially in low light.

   (c) *Color Blindness and Red-Green Sensitivity*: Many people with color vision deficiencies (e.g., red-green color blindness) have difficulty distinguishing between red and green hues. This limitation occurs because the HVS has a specific arrangement of photoreceptors that can sometimes result in reduced sensitivity or overlap between red and green wavelengths.

   (d) *Color Use in Data Visualization*: In data visualization, green and yellow shades are often used to make certain elements stand out, as these colors are highly visible to most people. Conversely, blue and red tones may be used for background or secondary information because they are less distracting.

4. **Visual Masking**: In complex textures or high-frequency regions, the HVS masks certain distortions, making them less visible. IQA models use visual masking techniques to emphasize areas where distortions are more apparent to the human eye and downplay others. Here are some examples:

   (a) *High-Detail Backgrounds Masking Compression Artifacts*: In areas of an image with a lot of texture, such as grass, sand or fabric patterns, compression artifacts are harder to detect because the complexity of the texture "masks" these minor imperfections.

In contrast, compression artifacts would be much more visible in smooth, low-detail areas such as clear skies or solid-colored walls.

(b) *Textured vs. Smooth Areas in Movies and Video Games*: In video content, details like grain or noise are often added to scenes with simple backgrounds to make compression artifacts less noticeable. For instance, in a video game scene with foggy skies, adding a little grain helps mask any potential banding artifacts (visible steps in gradients), making the scene appear smoother.

(c) *Peripheral Vision in High-Resolution Displays*: In high-resolution displays, visual masking is utilized by reducing resolution or detail in peripheral areas where the HVS is less sensitive. This technique allows resources to focus on the central vision area, masking lower-quality elements in peripheral vision.

(d) *Patterned Clothes Masking Stains*: In everyday life, patterns or textures on clothing make small stains or wrinkles less noticeable. Similarly, in images, complex textures mask small visual inconsistencies, making flaws less apparent to the eye.

5. **Brightness and Adaptation**: The HVS adapts to different lighting conditions and brightness, influencing how details and contrast are perceived under various levels of light. An effective IQA model should simulate these adaptations to handle images viewed under diverse lighting environments. Here are some examples illustrating this property:

(a) *Transition from Bright Outdoors to Dim Indoors*: When you walk from a sunny outdoor environment into a dimly lit room, your eyes take a moment to adjust. Initially, it's hard to see in the darker setting, but after a short period, your vision adapts, and you can distinguish objects more clearly. This adaptation to lower light levels helps the HVS maintain effective vision across varying environments.

(b) *Night Driving*: When driving at night, oncoming headlights can cause temporary blindness because they are much brighter than the surroundings. However, after the headlights pass, the HVS quickly readjusts to the darker road. This rapid brightness adaptation is essential for safe driving in low-light conditions, though extreme contrasts can momentarily disrupt perception.

(c) *Use of HDR in Photography and Displays*: High Dynamic Range (HDR) technology in photography and displays is designed to replicate the HVS's adaptation capabilities. HDR images capture both very bright and dark areas, ensuring that details in both regions remain visible, closely simulating the human eye's ability to perceive a wide range of brightness levels in a single scene.

(d) *Dark Adaptation in Astronomy*: When astronomers observe stars, they avoid bright lights beforehand so their eyes can adjust to the dark (a process called "dark adaptation"). After about 20 minutes, the HVS becomes more sensitive to low light, making it easier to see faint stars. If exposed to bright light, even briefly, the eyes would lose this sensitivity.

6. **Temporal and Motion Sensitivity**: For video IQA, the HVS's sensitivity to motion, changes and temporal continuity is crucial. Distortions in static regions are more noticeable than those in motion, and IQA models should integrate temporal masking for realistic assessments in videos. Here are examples illustrating this:

(a) *Tracking Moving Objects*: The HVS can track objects in motion, such as a car passing by or a ball in a sports game, and adjust quickly to keep the object in focus. This is why you can clearly see and follow a moving object with little to no motion blur when your eyes are focused on it, even as it moves quickly across your field of vision.

(b) *Perception of Flicker in Screens*: Temporal sensitivity allows the HVS to detect flickering lights or screens, such as old CRT monitors or fluorescent lights that refresh at lower frequencies. When the refresh rate is too low (e.g., below 60 Hz), humans can perceive flicker, which may cause discomfort. Modern screens typically use higher refresh rates to avoid this flicker perception, aligning with the HVS's high temporal sensitivity.

(c) *Motion Blur in Fast-Paced Video*: The HVS is more tolerant of motion blur in fast-moving scenes, such as during action scenes in movies or sports broadcasts, because we are less sensitive to fine details when objects are moving quickly. IQA models and video codecs leverage this by allowing more compression and lower detail in fast-moving frames, which appear natural due to the HVS's temporal sensitivity.

(d) *Peripheral Vision Sensitivity to Movement*: The HVS is highly sensitive to movement in peripheral vision, an evolutionary trait that helps detect potential threats or changes in the environment. For instance, even a small movement seen out of the corner of the eye, such as a bird flying or leaves rustling, can grab attention, showcasing the HVS's heightened motion sensitivity in the periphery.

(e) *Motion Artifacts in Compressed Video*: In the highly compressed video, temporal sensitivity means that artifacts (such as blocky distortions) are more noticeable in static scenes than in scenes with movement. Compression techniques often apply more aggressive data reduction in fast-moving scenes, knowing that the HVS will be less likely to detect artifacts due to reduced temporal sensitivity during motion.

7. **Structural Perception**: Humans are sensitive to structural elements, as they provide context and familiarity. Structural similarity metrics, such as those used in SSIM (Structural Similarity Index), are designed to capture how the HVS perceives overall image structures. Here are examples illustrating structural perception:

(a) *Face Recognition*: The HVS is highly attuned to the structure of human faces. Even minor structural changes, such as the position of the eyes, nose or mouth, can impact recognition. This sensitivity to facial structure allows us to distinguish between familiar and unfamiliar faces almost instantly. Image quality assessments often use metrics such as SSIM (Structural Similarity Index) to model how humans perceive similarity by focusing on structural details.

(b) *Text Readability*: Our eyes are sensitive to the structure of text, which allows us to recognize letters and words quickly. Minor variations in font style and structure still enable readability, but if letters are distorted too much, they become difficult to recognize. Structural perception here involves distinguishing familiar shapes and arrangements, facilitating fast reading even in different fonts and styles.

(c) *Perceiving Objects in Complex Scenes*: In a cluttered environment, the HVS can quickly pick out familiar structures, such as a chair or a car, due to its ability to recognize basic structural forms despite background noise. This is particularly useful in everyday life, like finding a friend in a crowd or locating a specific item in a busy room.

(d) *Pattern Recognition in Nature*: Structural perception helps us recognize natural patterns, such as identifying leaves on a tree or spotting animals camouflaged in their surroundings. The HVS is attuned to specific structural patterns, making it easier to identify familiar objects even in complex visual scenes.

(e) *Compression in Images and Videos*: Structural perception allows the HVS to identify important structures and ignore minor artifacts in digital media. Compression algorithms often prioritize retaining structural information while discarding redundant details, as humans are more likely to notice missing or distorted structures than minor color or brightness changes.

## 2.3 HVS-based Image Quality Assessment Methods

Human Visual System (HVS)-based image quality assessment methods evaluate image quality by mimicking how humans perceive visual information. These methods include visual saliency models, which prioritize regions likely to attract attention; perceptual metrics that align with human perception, such as the Just Noticeable Difference (JND), and hybrid approaches such as Bionics-inspired SSIM, which combines saliency with structural assessment for a more accurate and human-like quality evaluation. By leveraging principles of human vision, these methods provide a comprehensive and realistic assessment of image quality.

### 2.3.1 *Bionics methods*

These methods are inspired by how the human visual system (HVS) perceives and processes images. They aim to replicate the complexities of human vision to assess image quality more accurately. Key characteristics include the following:

1. **Visual Physiology and Psychophysics:** These methods often incorporate models that mimic the human eye's response to various visual stimuli. This includes how the eye perceives color, brightness, contrast and spatial frequency.
2. **Perceptual Metrics:** They use perceptual metrics that are designed to align with human visual perception. For example, methods might consider the Just Noticeable Difference (JND) to quantify how much a change in the image needs to be for it to be perceptible to the human eye.

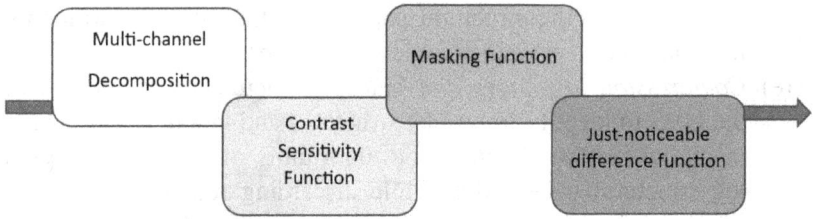

**Figure 2.3.**    Process of bionics method [24].

3. **Attention Models:** These methods may also integrate attention models
   that predict which parts of an image are most likely to attract human
   attention, leading to a more nuanced quality assessment.

Bionics method is the bottom-up approach that builds image quality
assessment metric based on modeling of some HVS concepts obtained by
physiology and psychophysics experiments, such as just-noticeable difference (JND) function, contrast sensitivity function, masking function and
multi-channel decomposition.

The process of the bionics methods is shown in Figure 2.3. In multi-
channel decomposition such as multi-scale geometric transform [25], and
Cortex transform [26], multiple filters such as localized, band-pass and
oriented filters are used to estimate image quality. Techniques proposed
in Ref. [27] have a model contrast sensitivity function using a band-pass
filter with multiple spatial frequencies to estimate quality. The masking
function explains that the visibility of one stimulus is affected due to
the presence of another. Several different types of masking methods have
been proposed in the literature. Daly visible differences predictor (VDP)
proposed in Ref. [28] has used Watson's cortex transform for the channel decomposition of reference and distorted images. Further, a difference map has been computed between decomposed distorted and reference
images that predict the image quality. The eye optics function is modeled
using a point-spread function (PSF) by Lubin [29]. This technique first
filters reference and distorted images using low-pass PSF and then dissim-
ilarities are computed between filtered images as image quality. Another
decomposition method is proposed in Ref. [30] using a quadrature mir-
ror filter. This filter decomposes an image into 16 sub-bands. Coefficients
obtained at each band are used to estimate quality of an image. Six orientations based on adopted steerable pyramid decomposition have been
presented [31] for quality assessment. DCT-based method has been used in

Ref. [32] to compute the perceptual error between distorted and reference images. This method has computed errors for every coefficient in every block of input images. Wavelet transform-based method is proposed in Ref. [33] that has estimated variance of wavelet coefficients of a distorted image to estimate image quality. Another method presented in Ref. [34] has used wavelet transform to quantify noise in an image. Spatial-temporal CSF model inspired by the excitation channel and inhibition channel has been developed in Ref. [35] to estimate quality. Spatio-temporal CIELAB method has been used in Ref. [36] to estimate video quality. Perceptual distortion metric (PDM) has been proposed in Ref. [37] for estimating video quality. This metric has used a steerable pyramid transform to decompose an image into multiple channels. Further, coefficients presented in each channel are weighted by contrast sensitivity function. Finally, Lp-norm summation method has been performed for the integration of the difference between coefficients of a distorted and a reference image. A technique proposed in Ref. [38] has utilized three HVS concepts, such as CSF, multiscale geometric analysis and Weber's law of just noticeable difference to estimate quality of an image.

## 2.3.2 *Engineering methods*

These methods rely on mathematical and statistical models to evaluate image quality. While they also aim to simulate aspects of the HVS, they often use more formal, technical approaches. Key characteristics include following:

1. **Feature Extraction:** These methods extract various features from the image, such as edges, textures and color information, and use these features to evaluate quality.
2. **Statistical Analysis:** They apply statistical techniques to analyze the extracted features. This might include calculating moments, entropy, correlation coefficients and other statistical measures.
3. **Modeling Distortions:** Engineering methods often model specific types of distortions (e.g., blur, noise and compression artifacts) and assess how these distortions impact image quality.

Engineering method is concerned only with the input and output of the HVS, i.e., only consider the input images and output image quality instead of its details. Figure 2.4 shows the framework of the engineering method.

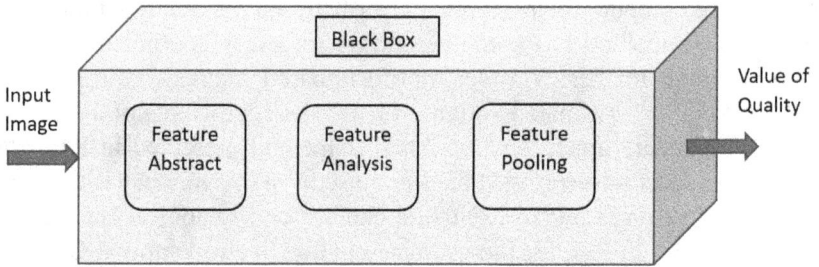

**Figure 2.4.**   Process of engineering method [24].

According to this category, image quality assessment technique is developed with prior knowledge of distortion types, similar to characteristics of HVS. These methods have low computational cost and good efficiency and therefore can be suitable in real-time applications. A technique named picture quality scale (PQS) proposed in Ref. [39] has exploited spatial frequency error weighted, luminance encoding error and random error to assess image quality. The fuzzy theory has been utilized in Ref. [40] to compare consistency and similarity between a distorted and a reference image. Sensitivity to structural information of HVS characteristics has been modeled into structural similarity (SSIM) image quality metric in technique [41]. SSIM-based image quality assessment method called a content-based metric (CBM) has been proposed in Ref. [42] where the output of SSIM (structural information) is divided by textures, edges and flat regions. An image quality assessment technique is presented in Ref. [43] that estimates quality using temporal variance, normalized color error and contour feature of the luminance image. Video quality metric (VQM) [44] method has used spatial and temporal distortion pooling to assess the quality of an image. There are some image quality assessment techniques found in the literature that are simple and effective, such as motion-estimated temporal consistency [45], discriminative local harmonic strength [46], temporal variations of spatial distortions [47], cross-dimensional perceptual quality assessment [48], depth map-based stereoscopic [49] and temporal trajectory aware quality assessment [50].

A technique proposed in Ref. [51] has estimated quality of an image by adapting various conditions of visual angles of edges and luminance. A technique for image quality assessment named visual information fidelity (VIF) has been presented in Ref. [52] that has modeled an image with a Gaussian mixture model in the wavelet domain and extracted features. A divisive normalization image representation and adaptive signal decomposition have been proposed in Refs. [53] and [54], respectively, to assess image quality. In order to simulate HVS, the JND model is transformed into a DCT domain to assess quality of an image. Another DCT domain transformation-based image quality assessment technique has been proposed in Ref. [55]. This method has computed visibility error by temporal fluctuations between a distorted image and a reference image. Techniques such as hybrid wavelets and directional filter banks (HWD) in Ref. [56], neural network in Ref. [57] and contourlet domain in Ref. [58] have been proposed to assess image quality.

If the characteristics of HVS are accurately modeled, then the precise estimation of image quality can be possible. However, there are some limitations due to the complexity of HVS. Therefore, to better understand HVS, there is a need for deep research in the field of physiology and psychology. The bionics method works by simulating HVS concepts and is more reasonable to develop. On the other hand, engineering methods deal with explaining some high-level cognitive concepts and the de-correlation of a natural image. However, they rely on mathematical models completely and modeling and parameter setting are more difficult than bionics method. In the future, it is very significant to fuse the concepts of bionics and engineering methods to build an efficient image quality assessment metric.

### 2.3.3 Combining both approaches: Bionics-inspired SSIM

Bionics-inspired SSIM (Structural Similarity Index) is inspired by how the human visual system perceives and processes images, focusing on areas that attract more attention. It integrates visual saliency models with SSIM to prioritize regions of the image that are more visually important.

The first step in the approach is to analyze the image and create a saliency map. This map highlights regions that are likely to attract human attention based on factors, such as color contrast, brightness and edge information. Various algorithms can be used to generate this map. For example, the Itti–Koch model uses features such as intensity, color and orientation to detect salient regions. These features are combined to form a comprehensive saliency map. The next step is SSIM calculations. The Structural Similarity Index (SSIM) evaluates image quality by comparing the structural information in an image, considering luminance, contrast and structure. SSIM values range from -1 to 1, where 1 indicates perfect similarity.

Bionics-inspired SSIM is a hybrid approach that combines the principles of the human visual system (HVS) with the traditional SSIM method to provide a more accurate and human-like assessment of image quality. It combines visual saliency models with SSIM. The saliency map identifies important regions in the image, and SSIM is applied more heavily to these regions.

In the bionics-inspired SSIM approach, the image is initially analyzed to create a saliency map that identifies regions likely to attract human attention. The SSIM is then applied, with calculations weighted according to the saliency map, thus giving more importance to these visually significant areas. So, instead of applying SSIM uniformly across the entire image, the calculations are weighted according to the saliency map. This means regions deemed important by the saliency map are given more weight during the SSIM calculation. This ensures that regions deemed important by the saliency map are prioritized during the quality assessment. The weighted SSIM scores from different regions are subsequently combined to produce an overall quality score, reflecting both structural similarity and perceptual importance. The result is an overall quality score that provides a more human-like assessment of image quality, considering the structure and the areas that are likely to attract human attention.

This hybrid method effectively balances the detailed structural assessment provided by SSIM with the attention-guided evaluation derived from visual saliency, resulting in a more accurate and human-like quality assessment. It considers both the structural integrity of the image and the perceptual significance of different regions, providing a comprehensive evaluation that closely mimics human visual perception.

**Benefits of the Hybrid Approach:** The bionics-inspired SSIM method effectively combines the strengths of bionics methods and engineering methods to provide a comprehensive and human-like image quality assessment. By considering both structural integrity and perceptual importance, it offers a more accurate and reliable evaluation of image quality:

1. **Accuracy:** By integrating visual saliency into the SSIM calculation, the assessment aligns more closely with how humans perceive image quality. This results in a more accurate evaluation.
2. **Human-Like Perception:** The approach mimics the human visual system by giving more importance to regions that attract attention, leading to a quality score that better reflects human perception.
3. **Enhanced Evaluation:** This method provides a balanced assessment, capturing both the detailed structural information and the perceptual significance of different regions.

# References

[1]   D. Marr, *Vision: A Computational Investigation into the Human Representation and Processing of Visual Information.* MIT press, Cambridge, 2010.

[2]   A. B. Wandell, *Foundations of Vision.* MA: Sinauer Associates, Sunderland, MA, 1995.

[3]   R. N. Carlson, *Foundations of Physiological Psychology (6th Edition).* Allyn & Bacon, Boston, 2004.

[4]   Thomas T. Norton, David A. Corliss, and James E. Bailey, *Psychophysical Measurement of Visual Function.* Butterworth-Heinemann, Boston, 2002.

[5]   W. R. Rodieck, "The primate retina," *Neurosciences*, 4, 203–278, 1998.

[6]   G. Westheimer, The eye as an optical instrument. *Handbook of Human Perception and Performance*, Wiley and sons, New York, 1986.

[7]   T. D. Shou, *Brain Mechanisms of Visual Information Processing.* Shanghai Science and Education Press, China, 1997.

[8]   Z. Li, "A neural model of contour integration in the primary visual cortex," *Neural Comput.*, 10(4), 903–940, 1998.

[9]   Y. G. G. R. Peli E, Arend LE, "Contrast sensitivity to patch stimuli: effects of spatial bandwidth and temporal presentation," *Spatial Vision*, 7(1), 1–14, 1993.

[10]  M. A. Treisman and G. Gelade, "A feature-integration theory of attention," *Cognitive Psychology*, 12(1), 97–136, 1980.

[11]  G. E. Legge and J. M. Foley, "Contrast masking in human vision," *Journal of the Optical Society of America*, 70(12), 1458–1471, 1980.

[12]  A. A. Michelson, *Studies in Optics.* IL: Univ. Chicago Press, Chicago, 1927.

[13]  R. J. Campbell FW, "Application of fourier analysis to the visibility of gratings," *Journal of Psychology*, 197(3), 551–566, 1968.

[14]  A. B. Watson, R. Borthwick, and M. Taylor, "Image quality and entropy masking," in *Human Vision and Electronic Imaging II*, 3016.SPIE, pp. 2–12, 1997.

[15]  S. A. Klein, T. Carney, L. Barghout-Stein, and C. W. Tyler, "Seven models of masking," in *Human Vision and Electronic Imaging II*, 3016. SPIE, 13–24, 1997.

[16]  E. DeYoe and D. Van Essen, "Concurrent processing streams in monkey visual cortex," *Trends in Neurosciences*, 11(5), 219–226, 1988.

[17]  W. James, *The Principles of Psychology.* H. Holt, New York, 1890.

[18]  L. Itti, C. Koch, and E. Niebur, "A model of saliency-based visual attention for rapid scene analysis," *IEEE Transactions on Pattern Analysis and Machine Intelligence*, 20(11), 1254–1259, 1998.

[19]  J. Harel, C. Koch, and P. Perona, "Graph-based visual saliency," in *Advances in Neural Information Processing Systems*, 19, 2006.

[20]    U. Rajashekar, I. van der Linde, A. C. Bovik, and L. K. Cormack, "Gaffe: A gaze-attentive fixation finding engine," *IEEE Transactions on Image Processing*, 17(4), 564–573, 2008.

[21]    F. Crick, "Function of the thalamic reticular complex: The searchlight hypothesis." *Proceedings of the National Academy of Sciences*, 81(14), 4586–4590, 1984.

[22]    J. K. Tsotsos, "Analyzing vision at the complexity level," *Behavioral and Brain Sciences*, 13(3), 423–445, 1990.

[23]    X. Gao, W. Lu, D. Tao, and X. Li, "Image quality assessment and human visual system," in *Proceedings of Visual Communications and Image Processing 2010*. International Society for Optics and Photonics, 2010, 77440Z–77440Z.

[24]    Y. Ding, *Image Quality Assessment Based on Human Visual System Properties*. Springer Berlin Heidelberg, 2018, 63–106.

[25]    J. K. Romberg, M. B. Wakin, and R. G. Baraniuk, "Multiscale geometric image processing," in *Visual Communications and Image Processing*, 5150. International Society for Optics and Photonics, 2003, 1265–1272.

[26]    A. B. Watson, "The cortex transform: Rapid computation of simulated neural images," *Comput. Vision Graph. Image Process.* 39(3), 311–327, 1987.

[27]    M. KT, "The contrast sensitivity of human colour vision to red-green and blue-yellow chromatic gratings," *Journal of Physiology*, vol. 359, pp. 381–400, 1985.

[28]    S. J. Daly, "Visible differences predictor: an algorithm for the assessment of image fidelity," in B. E. Rogowitz, (Ed.) *Human Vision, Visual Processing, and Digital Display III*, International Society for Optics and Photonics, 1992, pp. 2–15.

[29]    J. Lubin, "A visual discrimination model for imaging system design and evaluation," in *World Scientific*, 1995, pp. 207–220.

[30]    R. J. Safranek and J. D. Johnston, "A perceptually tuned sub-band image coder with image dependent quantization and post-quantization data compression," *International Conference on Acoustics, Speech, and Signal Processing*, 3, 1945–1948, 1989.

[31]    P. C. Teo and D. J. Heeger, "Perceptual image distortion," in B. E. Rogowitz and J. P. Allebach (Eds.) *Human Vision, Visual Processing, and Digital Display V*, International Society for Optics and Photonics, 1994, pp. 127–141.

[32]    A. B. Watson, "DCT quantization matrices visually optimized for individual images," in *Human Vision, Visual Processing, and Digital Display IV*, 1913.International Society for Optics and Photonics, 1993, 202–216.

[33]　W. Lu, X. Gao, D. Tao, and X. Li, "A wavelet-based image quality assessment method," *International Journal of Wavelets, Multiresolution and Information Processing*, 06(04), 541–551, 2008.

[34]　A. Watson, G. Yang, J. Solomon, and J. Villasenor, "Visibility of wavelet quantization noise," *IEEE Transactions on Image Processing*, 6(8), 1164–1175, 1997.

[35]　F. Lukas and Z. Budrikis, "Picture quality prediction based on a visual model," *IEEE Transactions on Communications*, 30(7), 1679–1692, 1982.

[36]　X. Tong, D. J. Heeger, and C. J. V. den Branden Lambrecht, "Video quality evaluation using ST-CIELAB," in B. E. Rogowitz and T. N. Pappas (Eds.) *Human Vision and Electronic Imaging IV*, International Society for Optics and Photonics, 1999, pp. 185–196.

[37]　S. Winkler, "Perceptual distortion metric for digital color video," in B. E. Rogowitz and T. N. Pappas (Eds.) *Human Vision and Electronic Imaging IV*, 3644. International Society for Optics and Photonics, 1999, pp. 175–184.

[38]　X. Gao, W. Lu, X. Li, and D. Tao, "Wavelet-based contourlet in quality evaluation of digital images," *Neurocomputing*, 72(1), 378–385, 2008.

[39]　M. Miyahara, K. Kotani, and V. Algazi, "Objective picture quality scale (PQS) for image coding," *IEEE Transactions on Communications*, 46(9), 1215–1226, 1998.

[40]　D. Van der Weken, M. Nachtegael, and E. E. Kerre, "Using similarity measures and homogeneity for the comparison of images," *Image and Vision Computing*, 22(9), 695–702, 2004.

[41]　Z. Wang, A. Bovik, H. Sheikh, and E. Simoncelli, "Image quality assessment: from error visibility to structural similarity," *IEEE Transactions on Image Processing,*, 13(4), 600–612, 2004.

[42]　X. Gao, T. Wang, and J. Li, "A content-based image quality metric," in *Rough Sets, Fuzzy Sets, Data Mining, and Granular Computing*. Springer Berlin Heidelberg, 2005.

[43]　A. Hekstra, J. Beerends, D. Ledermann, F. de Caluwe, S. Kohler, R. Koenen, S. Rihs, M. Ehrsam, and D. Schlauss, "Pvqm — a perceptual video quality measure," *Signal Processing: Image Communication*, 17(10), 781–798, 2002.

[44]　S. Wolf and M. H. Pinson, "Spatial-temporal distortion metric for in-service quality monitoring of any digital video system," in A. G. Tescher, B. Vasudev, V. M. B. Jr., and B. Derryberry (Eds.) *Multimedia Systems and Applications II*, International Society for Optics and Photonics, 1999, pp. 266–277.

[45]  P. V. Pahalawatta and A. M. Tourapis, "Motion estimated temporal consistency metrics for objective video quality assessment," in *International Workshop on Quality of Multimedia Experience*, 2009, pp. 174–179.

[46]  I. P. Gunawan and M. Ghanbari, "Reduced-reference video quality assessment using discriminative local harmonic strength with motion consideration," *IEEE Transactions on Circuits and Systems for Video Technology*, 18(1), 71–83, 2008.

[47]  A. Ninassi, O. Le Meur, P. Le Callet, and D. Barba, "Considering temporal variations of spatial visual distortions in video quality assessment," *IEEE Journal of Selected Topics in Signal Processing*, 3(2), 253–265, 2009.

[48]  G. Zhai, J. Cai, W. Lin, X. Yang, W. Zhang, and M. Etoh, "Cross-dimensional perceptual quality assessment for low bit-rate videos," *IEEE Transactions on Multimedia*, 10(7), 1316–1324, 2008.

[49]  C. T. E. R. Hewage, S. T. Worrall, S. Dogan, S. Villette, and A. M. Kondoz, "Quality evaluation of color plus depth map-based stereoscopic video," *IEEE Journal of Selected Topics in Signal Processing*, 3(2), 304–318, 2009.

[50]  M. Barkowsky, J. Bialkowski, B. Eskofier, R. Bitto, and A. Kaup, "Temporal trajectory aware video quality measure," *IEEE Journal of Selected Topics in Signal Processing*, 3(2), 266–279, 2009.

[51]  D. M. Chandler and S. S. Hemami, "VSNR: A wavelet-based visual signal-to-noise ratio for natural images," *IEEE Transactions on Image Processing*, 16(9), 2284–2298, 2007.

[52]  H. Sheikh, A. Bovik, and G. de Veciana, "An information fidelity criterion for image quality assessment using natural scene statistics," *IEEE Transactions on Image Processing*, 14(12), 2117–2128, 2005.

[53]  Q. Li and Z. Wang, "Reduced-reference image quality assessment using divisive normalization-based image representation," *IEEE Journal of Selected Topics in Signal Processing*, 3(2), 202–211, 2009.

[54]  U. Rajashekar, Z. Wang, and E. P. Simoncelli, "Quantifying color image distortions based on adaptive spatio-chromatic signal decompositions," in *16th IEEE International Conference on Image Processing (ICIP)*, 2009, pp. 2213–2216.

[55]  V. Baroncini and A. Pierotti, "Single-ended objective quality assessment of DTV," in A. G. Tescher, B. Vasudev, V. M. B. Jr., and B. Derryberry (Eds.) *Multimedia Systems and Applications II*, International Society for Optics and Photonics, 1999, pp. 244–253.

[56]  X. Li, D. Tao, X. Gao, and W. Lu, "A natural image quality evaluation metric," *Signal Processing*, 89(4), 548–555, 2009.

[57]  P. Gastaldo, R. Zunino, and S. Rovetta, "Objective assessment of MPEG-2 video quality," *Journal of Electronic Imaging*, 11, 365–374, 2002.

[58]  D. Tao, X. Li, W. Lu, and X. Gao, "Reduced-reference iqa in contourlet domain," *IEEE Transactions on Systems, Man, and Cybernetics, Part B (Cybernetics)*, 39(6), 1623–1627, 2009.

# Chapter 3

# Image Quality Factors

Quality considers the significant attributes (quality factors) present in an image and their weighted combination. It is how accurately an image is processed by imaging systems such as capturing, storing, compressing, transmitting and displaying without affecting quality attributes. Image quality describes signal processing characteristics and depicts the pleasing perceptual quality present in an image for human viewers. Let's explore more on the degradation of an image to understand image quality better. To better get an insight into the image quality, we explain the factors responsible for degradation of an image, such as sharpness, noise, distortions, contrast and illumination. They are discussed in the following subsections.

## 3.1 Sharpness

Sharpness is the significant quality factor that depicts the clarity of detail present in an image. The sharpness in image quality can be seen in high-resolution photography. In a sharp image, edges and fine details are clearly defined, such as individual leaves on a tree or texture on fabric. Sharpness enhances the clarity and distinction between elements, making the image look crisper. Conversely, an unsharp image may appear blurry, with details blending into each other, which can happen due to camera shake, out-of-focus lenses or insufficient resolution for the subject detail.

An example of reduced sharpness is shown in Figure 3.1. An original image is filtered with a Gaussian filter which is set to 0.7 sigma to reduce the sharpness. The sharpness of an image can be affected by the lens

**Original          |          Blurred**

**Figure 3.1.**   An example of the reduction in sharpness [1].

(focal length, aperture, design and manufacturing quality), sensors and image processing steps, such as noise reduction and sharpening. Moreover, sharpness can also be disturbed by atmospheric disturbances, focus accuracy and camera shake.

Sharpening of an image is used to restore lost sharpness. Figure 3.2 shows an example of enhancement by sharpening an image. A widely used method named adaptive bilateral filter (ABF) is proposed in Ref. [2] to sharpen an image by increasing the slope of the edges. However, sharpening has some limitations. It cannot recover the lost sharpness where modulation transfer function (MTF, also called spatial frequency response (SFR)) is obtained with a very low value, i.e., less than 10%. It is observed that halos have appeared near contrast boundaries when an image is over-sharpened at large magnifications (see Figure 3.3).

(a) Scanned text image

(b) Sharpened by ABF method [2]

**Figure 3.2.**   An example of image sharpening.

Original          |          Oversharpened

**Figure 3.3.**   An example of an over-sharpened image [1].

## 3.2   Noise

Noise is defined as random variations in pixels in an image. It is caused due to the basic physics effect, i.e., thermal energy of heat and the nature of light. For example, in a dimly lit photo of a night scene, you might note tiny specks or graininess across smooth surfaces such as the sky or walls. This noise disrupts the image's clarity, making areas that should look smooth appear textured or pixelated. High levels of noise can obscure fine details, reducing the overall quality and visual appeal of the image. Figure 3.4 shows an example of a noisy image. The following factors are directly associated with noise:

1. **Pixel size**: To get a better signal-to-noise ratio (SNR) (the better the signal quality), a larger pixel size is required (i.e., more photons). The

Original                    |                    **Added noise**

**Figure 3.4.**   An example of noisy image [1].

number of photo-electrons generated by the photons is proportional to the sensor area. If we increase the size of a pixel, we need to increase the size of the sensor area. It generates more photo-electrons that lead to better SNR to obtain a good quality signal.

2. **ISO speed (Exposure Index) setting**: ISO speed is controlled by amplifying the signal in digital cameras. Higher ISO speed may produce more noise in the images. Therefore, for proper photography (to get less noise), ISO speed must be tested with variations.

3. **Exposure time**: Short exposures with bright light lead to fewer noisy images than long exposures with dim light. Exposure time must be tested with variations for generating good-quality images.

Original

(a)

Barrel Distortion

(b)

Pincushion Distortion

(c)

**Figure 3.5.**    An example of barrel and pincushion distorted images [3].

4. **Digital processing**: Generally, sensors have 12-bit analogue-to-digital (A-to-D) converters, and they may not generate noise while digitization at the sensor level. But, noise may be produced when an image is an 8-bit (24-bit color) JPEG.

# 3.3 Lens (optical) Distortion

Its when a clear straight line, in reality, becomes curved when photographed with a flawed lens. This kind of distortion creates trouble in photogrammetry and architectural photography applications where measurement is taken from the images. It is noted that the third-order equation $r_u = r_d + k r_d^3$ is used to approximate lens distortion, where $r_u$ is the undistorted and $r_d$ is the distorted radius. The sign of $k$ decides the type of distortion, i.e., barrel or pincushion (shown in Figure 3.5). There is another distortion called mustache distortion that can be approximated by the fifth-order equation $(r_u = r_d + h_1 r_d^3 + h_2 r_d^5)$. Heres a closer look at the main types:

1. **Barrel Distortion**: Lines curve outward from the center, giving a "barrel" shape. Common in wide-angle lenses, it's most visible near the edges of the image. Ideal for landscapes but distorts architecture.
2. **Pincushion Distortion**: Lines curve inward toward the center, "pinching" the image. Found in telephoto lenses, it's often seen in portrait shots where edges appear pulled in.
3. **Mustache (Complex) Distortion**: This combines barrel and pincushion effects, with lines bulging outward near the center and curving inward at the edges. Often seen in zoom lenses, it's complex to correct but can be useful for dynamic scenes.

# 3.4 Contrast

The difference between maximum and minimum intensity of pixels in an image is termed contrast. In other words, contrast helps make an object distinguishable by the difference in luminance or color. An example of contrast in an image is shown in Figure 3.6. We, humans, can perceive objects similarly regardless of variation in illumination because HVS is more sensitive to contrast than luminance. It is noted that increasing darkness will increase contrast in bright areas and decrease contrast in an image, while increasing the brightness will have the opposite effect.

There are various definitions of contrast are available. Luminance contrast can be defined as

$$\frac{Luminance\ difference}{Average\ luminance} \tag{3.1}$$

**Figure 3.6.** The contrast in the left half of the image is lower than that in the right half [4].

The above equation depicts that a small luminance difference is significant if the average luminance is low and a small luminance difference is avoided if the average luminance is high. Another contrast is known as Weber contrast and it is defined as

$$\frac{I - I_b}{I_b} \tag{3.2}$$

where $I$ and $I_b$ are the luminance of the image's features and the image's background, respectively. The Weber contrast can be useful where features are present in a small area with a large uniform background. In other words, it is useful where background luminance is almost equal to average luminance.

When both dark and bright areas in an image are approximately equal, Michelson contrast is a useful metric. The Michelson contrast is defined as

$$\frac{I_{max} - I_{min}}{I_{max} + I_{min}} \tag{3.3}$$

where $I_{max}$ and $I_{min}$ represent the highest and lowest luminance. Root mean square (RMS) contrast is a measure of the standard deviation of intensity

(a)  Image with low contrast

(b)  Histogram Equalized image

**Figure 3.7.**  An example of low-contrast image enhancement [5].

of pixels in an image. It is given as follows:

$$\sqrt{\frac{1}{MN} \sum_{i=0}^{N-1} \sum_{j=0}^{M-1} (I_{ij} - \bar{I})^2} \qquad (3.4)$$

where intensity $I_{ij}$ is the $i$-th $j$-th location in the image of size $M$ by $N$. $\bar{I}$ is the average intensity of all pixels.

An example of a poor contrast image and its enhancement is shown in Figure 3.7. As we can see, Figure 3.7 (a) is a low-contrast image having poor visual quality. Figure 3.7 (b) is the enhanced image (enhanced using a well-known method called Histogram Equalization) having better visual quality.

## 3.5  Illumination

The light you observe on your screen is the illumination. In daily life, we can see that an image captured in low light (poor illumination) contains less information than an image captured in good lighting conditions (good illumination). Illumination plays an important part in image formation. An image is formed by two physical variables, i.e., object's reflectance and illumination (under which a scene is viewed), and it is defined as

$$I = R * L \qquad (3.5)$$

This equation depicts that light with spectral power distribution $L$ falls on a surface that reflects light on a per wavelength basis according to its reflectance function $R$. It is clear that illumination is a significant factor in image formation and, independent of it, can degrade image quality. An example of a natural image with poor illumination and its enhancement by method [6] is shown in Figure 3.8.

Light falls on a surface with a specific angle can change the result. For example, front illumination or backlighting, direct or diffuse illumination, and bright-field or dark-field illumination are several ways to throw light at a specific angle. Figure 3.9 explicates how different an object may appear depending on how the illumination is organized. The different organizations of illumination are briefly explained as follows:

1. In direct front illumination, illumination is parallel to the optical axis of the camera and produces non-uniform images. Figure 3.9 (a) shows the illumination organization and a generated image.

(a) Image with poor illumination

(b) Image enhancement using method [6]

**Figure 3.8.**   An example of poor illumination enhancement.

2. Figure 3.9 (b) presents an example of diffuse bright-field illumination. We can see the high contrast between the background and the object. Therefore, the output image appears uniform. However, there are fewer details observed in this illumination organization.

3. Diffuse dark-field illumination occurs when light falls with an oblique angle from a ring. Figure 3.9 (c) shows this organization that generated good details in the output image.

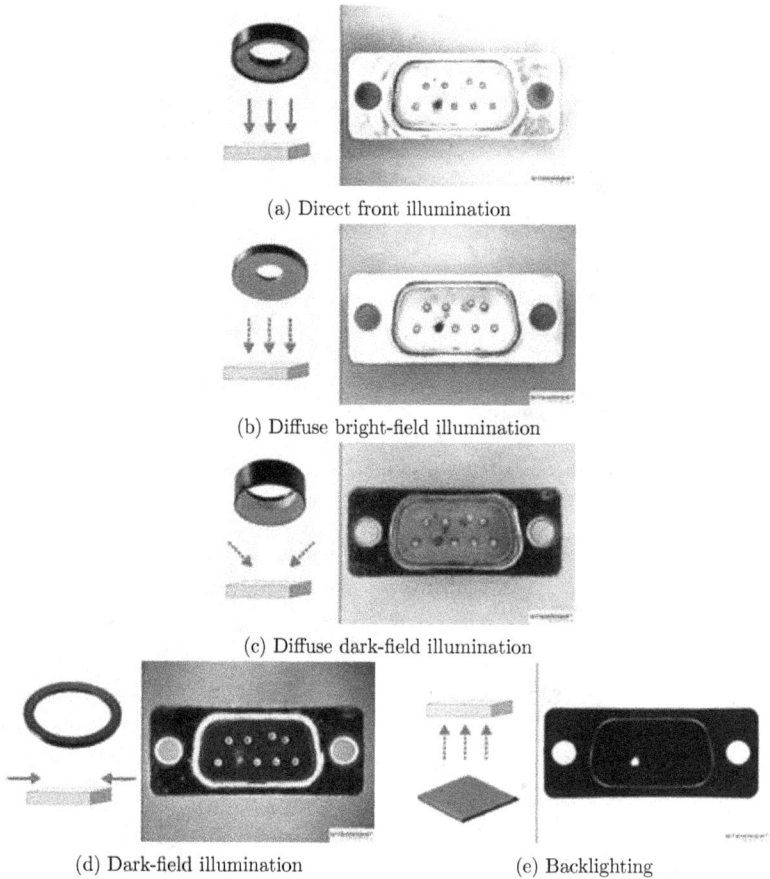

(a) Direct front illumination

(b) Diffuse bright-field illumination

(c) Diffuse dark-field illumination

(d) Dark-field illumination          (e) Backlighting

**Figure 3.9.**   Example of different illumination conditions [7].

4. Dark-field illumination is produced when a shallow angle of light falls on the object. In Figure 3.9 (d), we can see that the top parts of the object, such as pins and connectors, are visibly bright and can be easily detected by image analysis for further application.
5. Figure 3.9 (e) shows an example of backlighting. Here, light is coming from the rear of an object and it penetrates only where an obstruct is not present. This organization can be useful in detecting holes in an object.

All the above-mentioned factors are responsible for the degradation in image quality. Therefore, these factors should be carefully estimated and then enhanced to obtain a good-quality image.

# References

[1]  Image quality factors. Accessed 07-04-2023, Available: https://www.imatest.com/docs/iqfactors/1.

[2]  B. Zhang and J. P. Allebach, "Adaptive bilateral filter for sharpness enhancement and noise removal," *IEEE Transactions on Image Processing*, 17(5), 664–678, 2008.

[3]  J. Kelcey and A. Lucieer, "Sensor correction of a 6-band multispectral imaging sensor for UAV remote sensing," *Remote Sensing*, 4(5), 1462–1493, 2012.

[4]  Contrast (vision). Accessed 07-04-2023, Available: https://en.wikipedia.org/wiki/Contrast_%28vision%29.

[5]  Histogram equalization. Accessed 07-04-2023, Available: https://www.mathworks.com/help/images/image-import-and-export.html.

[6]  C. Guo, C. Li, J. Guo, C. C. Loy, J. Hou, S. Kwong, and C. Runmin, "Zero-reference deep curve estimation for low-light image enhancement," *CVPR*, 2020.

[7]  Different organisations of illumination. Accessed 07-04-2023, Available: https://www.stemmer-imaging.com/s/blog?language=en_US.

# Chapter 4

# Review of Image Quality Assessment Databases

In the last two decades, a significant number of image quality assessment databases with human-rated quality scores have come out. These databases have used a variety of subjective testing methodologies, viewing distances from screen, and ratings per image. Their benchmark quality ratings have different ranges and are either in the form of difference mean opinion scores (DMOS) or mean opinion scores (MOS). The content of a reference image is selected in an *ad hoc* manner. Further, multiple distorted images are simulated at different intensity levels of distortion and selected in an *ad hoc* manner. The goal of image quality assessment database is to generate distorted images such that the quality spectrum is uniformly represented. Mostly, image quality assessment databases consider single distortion per image which is not a practical approach as an image may be distorted by more than one distortion. We review databases which contain (1) color images, (2) natural images, (3) both reference and distorted images for the evaluation of FR-image quality assessment and (4) standard dynamic range (SDR) images. Tables 4.1 and 4.2 present a summary of image quality assessment databases [1]. Different databases are discussed as follows.

**Table 4.1.** Summary of LIVE and TID2013 image quality assessment databases.

| Database | Year | No. of ref. images | No. of dist. images | Dist. list (no. of images) | Dist. per image | Subjective test method | Subjective data type | Score range | Ratings per image | Viewing distance |
|---|---|---|---|---|---|---|---|---|---|---|
| LIVE R2 | 2006 | 29 | 779 | White Gaussian noise (145)<br>Gaussian blur (145)<br>JPEG compression (175)<br>JPEG compression (169)<br>Fast fading (145) | 1 | Single stimulus | DMOS | −2.64 to 111.47 | 23 | 2–2.5 screen heights |
| TID2013 | 2013 | 25 | 3000 | Additive Gaussian Noise (125)<br>Additive Noise in Color Components (125)<br>Spatially Correlated Noise (125)<br>Masked Noise (125)<br>High Frequency Noise (125)<br>Impulse Noise (125)<br>Quantization Noise (125)<br>Gaussian Blur (125)<br>JPEG Compression (125)<br>JPEG2000 Compression (125)<br>JPEG Transmission Errors (125)<br>JPEG2000 Transmission Errors (125)<br>Non Eccentricity Pattern Noise (125)<br>Local Block-wise Distortions (125)<br>Mean Shift (125)<br>Contrast Change (125)<br>Change of Color Saturation (125)<br>Multiplicative Gaussian Noise (125)<br>Comfort Noise (125)<br>Lossy Compression of Noisy (125) | 1–2 | pair-wise comparison | MOS | 0.24–7.21 | 30 | Varying |

**Table 4.2.** Some other important databases used in analysis of image quality assessment techniques.

| Database | Year | No. of ref. images | No. of dist. images | Dist. list (no. of images) | Dist. per image | Subjective test method | Subjective data type | Score range | Ratings per image | Viewing distance |
|---|---|---|---|---|---|---|---|---|---|---|
| CSIQ | 2010 | 30 | 886 | White Gaussian noise (150) Gaussian blur (150) JPEG compression (150) JPEG compression (150) Additive pink Gaussian noise (150) Global contrast decrements (116) | 1 | Simultaneous comparison | DMOS | 0–1 | 6 | 70 cm |
| VCLFER | 2012 | 23 | 552 | White Gaussian noise (138) Gaussian blur (138) JPEG compression (138) JPEG compression (138) | 1 | Single stimulus | MOS | 1.57–96.52 | 16–36 | – |
| CIDIQ | 2014 | 23 | 690 | Poisson noise (115) Gaussian blur (115) JPEG compression (115) JPEG compression (115) SGCK Gamut mapping (115) $\nabla$ E Gamut mapping (115) | 1 | Double stimulus | MOS | 1.8–7.65 1–7.76 | 17 | 50 cm 100 cm |
| LIVE MD | 2012 | 15 | 405 | White Gaussian noise (45) Gaussian blur (45) JPEG compression (45) Gaussian blur followed by JPEG compression (135) Gaussian blur followed by Gaussian noise (135) | 1 1 1 2 2 | Single stimulus | DMOS | 0.61–84.67 | 19 | Four screen heights |
| MDID 2013 | 2014 | 12 | 324 | Gaussian blur followed by JPEG compression followed by White Gaussian noise (324) | 3 | Single stimulus | DMOS | 0.32–0.55 | 25 | Four screen heights |
| MDID | 2017 | 20 | 1600 | Gaussian blur and/or contrast change followed by JPEG or JPEG compression followed by Gaussian noise | 1–4 | Pair comparison sorting | MOS | 0.08–7.92 | 33–35 | Two screen heights |
| MDIVL | 2017 | 10 | 750 | Gaussian blur followed by JPEG compression (350) Gaussian noise followed by JPEG compression (400) | 2 | Single stimulus | MOS | 1.41–97.97 | 12 | – |

## 4.1   Single Distortion Databases

In this category, each image is distorted with only one kind of distortion; that's why this category is referred to as singly distorted databases. There are many databases in the literature which fall into this category. A widely used LIVE Release 2 (LIVE R2) database [2, 3] is the best example of this category that is developed by the Laboratory for Image and Video Engineering at UT Austin. This database contains 29 reference images and 779 distorted images of a total of five distortions with an image resolution of $480 \times 720$ or up to $768 \times 512$. All reference images are shown in Figure 4.1.

This database has provided subjective ratings using stimulus methodology [6] where reference images also have been evaluated. A small training was provided to participants before taking subjective ratings from them. Participants have to provide a score in the range of 0–100 for all images. The database provides subjective data in the form of DMOS after outlier removal, where a lower DMOS represents better quality. Table 4.1 shows further details of this database.

An extension to TID2008 database [4] is presented in Ref. [7] named as Tampere Image Database 2013 (TID2013) database. In this database, 3000 distorted images were generated using 25 reference images having a resolution of $512 \times 384$. All reference images are shown in Figure 4.2. This database has 24 distortions with five distortion levels. For subjective ratings, from five countries, 971 subjects have been trained. Subjective scores have been taken using a tristimulus methodology [7] in two modes including remotely via the internet and laboratory environment. After outlier removal, MOS was obtained for the database, where higher MOS represents better quality. Refer to Table 4.1 for more details.

**Figure 4.1.**   All 29 reference images of LIVE database [3].

**Figure 4.2.** All reference images of TID database [4].

**Figure 4.3.** All reference images of CSIQ database [5].

A database proposed in Ref. [5] named the Computational and Subjective Image Quality (CSIQ) database has 30 reference images and 866 distorted images (a resolution of $512 \times 512$). All reference images are shown in Figure 4.3. It has six distortion types and four to five levels of distortion per type. For subjective ratings, distorted images and their corresponding reference images were displayed simultaneously. After outlier removal

DMOS was obtained, where a lower DMOS value represents better quality. For further details, see Table 4.2.

Another database named Video Communications Laboratory @ FER (VCLFER) database [8] contains 552 distorted images of 23 reference images. Distorted images were generated using four distortions with six levels per distortion. For subjective ratings, a single stimulus methodology was obtained, where a lower DMOS value represents better quality. For further details, see Table 4.2.

*Colourlab Image Database*: Image Quality (CID:IQ) [9] contains 690 distorted images of 23 reference images. Distorted images were generated using six distortions with five levels per distortion. The resolution of all images is $800 \times 800$. This database has provided two sets of MOS: one at 50 cm and another at 100 cm. A higher MOS value represents better visual quality.

## 4.2   Multiple Distortion Databases

These databases are referred to as multiply distorted databases and contain multiple distortions (two or more) per image; thereby, they are better at representing the practical situation of distortion. The first multiply distorted database is LIVE multiply distorted (LIVE MD) database [10]. Distorted images are generated by using 15 reference images, having a resolution of $1280 \times 720$. This database has 135 singly distorted images and 270 multiply distorted images. A total of three distortions (JPEG compression, Gaussian blur and white Gaussian noise) are considered in this database with three distortion levels. This database has used a combination of two distortions to generate multiply distorted images. One combination is Gaussian blur followed by JPEG compression and the other combination is Gaussian blur followed by white Gaussian noise. A small training was provided to participants before taking subjective ratings from them. Participants have to provide a score in the range of 0–100 for all images. The database provides subjective data in the form of DMOS after outlier removal, where a lower DMOS represents better quality. Table 4.2 shows further details of this database. Another database presented in Ref. [11] is Multiply Distorted Image Database 2013 (MDID2013) which has used JPEG compression, Gaussian blur and white Gaussian noise distortions. It has created 324 multiply distorted images using 12 reference images with 3 distortion levels. In this database, multiply distorted images have been generated by first undergoing blurring which is followed by JPEG compression followed by white noise. To get subjective ratings on a scale of 0–1, a single stimulus methodology [6] has been used. The database provides subjective data in the form of DMOS after outlier removal, where a lower DMOS represents better quality.

A database presented in Ref. [11] is Multiply Distorted Image Database (MDID) [12] (different from MDID2013) that has used Gaussian blur, Gaussian noise, JPEG, contrast change and JPEG2000 compression. It has created 324 multiply distorted images using 12 reference images with 4 distortion levels. This database has simulated distortions as follows. The distortion that may occur at acquisition time is simulated using contrast change and/or Gaussian blur. The distortion that may occur during image transmission is simulated by JPEG or JPEG2000 compression. At last, Gaussian noise is used to simulate display imperfections. This database has obeyed the following rules while adding distortions: (1) The final distorted image should have at least one distortion. (2) For compression, either JPEG or JPEG2000 is used. (3) There is no repetition of distortions. To get subjective ratings, the paired comparison sorting methodology [12] is used. The database provides subjective data in the form of MOS after outlier removal, where a higher MOS represents better quality.

A database called Multiple Distorted IVL (MDIVL) database presented in Ref. [13] has used 10 reference images to generate 750 multiply distorted images. To generate multiple distortions in an image, this database has obeyed the following combinations: (1) In Blur-JPEG combination, first, each reference image is blurred at seven levels using Gaussian blur, then five levels of JPEG compression are applied to each blurred image. (2) In Noise-JPEG combination, first, in each reference image, noise is added at ten levels using Gaussian noise, then four levels of JPEG compression are applied to each noisy image. To get subjective ratings, a single stimulus methodology [6] is used. Participants have to provide a score in the range of 0–100 for all images. The database provides subjective data in the form of MOS after outlier removal, where a higher MOS represents better quality. See Table 4.2 for more details.

## 4.3   Distortion Analysis of Above Presented Databases

It is important to know the distribution of distortion across the quality range for a database. It tells that the distortion distribution is uniform or

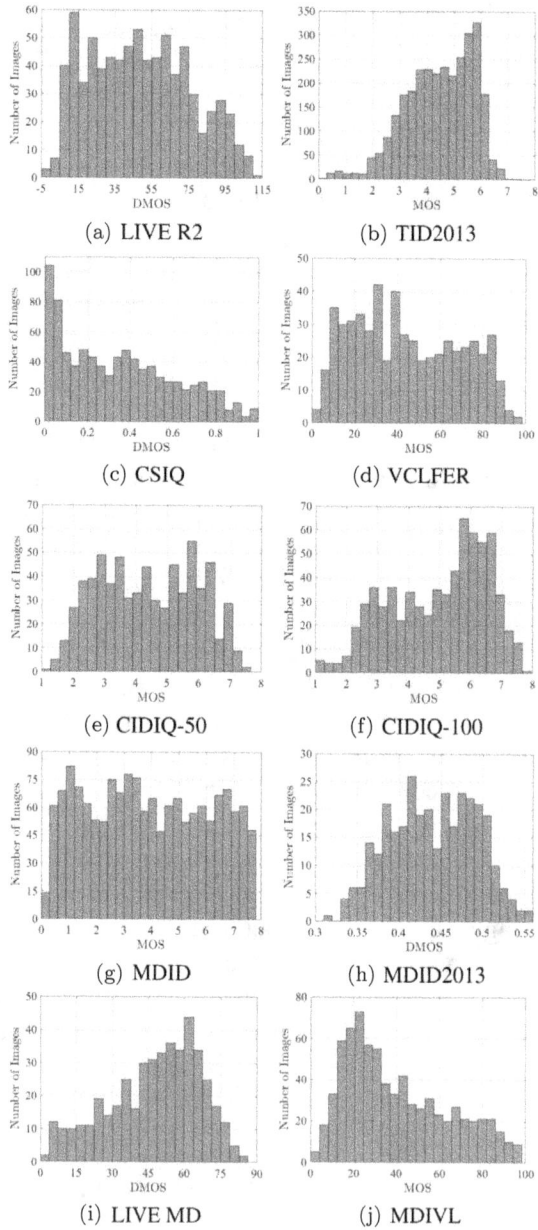

**Figure 4.4.** Histograms of MOS/DMOS of the nine image quality assessment databases. *Note*: CIDIQ database is formed by two viewing distances (50 cm and 100 cm).

non-uniform. In addition, it provides information about a database containing either a large number of better-quality images or more highly distorted images. Figure 4.4 shows the histogram of MOS/DMOS of the nine image quality assessment databases. A lower DMOS value signifies better visual quality, whereas a higher MOS value indicates better visual quality. We can see from the figure that the distortion distribution in MDID database is relatively uniform than other databases. Also, LIVE R2, VCLFER and CIDIQ-50 have a mild uniform distribution. TID2013 and CSIQ databases have more number of better-quality images. On the other hand, LIVE MD and MDIVL contain more low-quality images. It is presented in Ref. [7] that any image quality assessment technique found more difficult to estimate the quality of better-quality images than low-quality images. We can conclude that a database containing more number of better-quality images is more challenging than a database which contains more low-quality images.

## 4.4  Other Image Quality Assessment Databases

Apart from the databases mentioned in the above sections, there are a number of databases available in the literature. A database A57 presented in Ref. [14] has used three reference images to generate 54 distorted images. This database contains grayscale images of a resolution of $512 \times 512$. Six different distortions have been considered in this database including (1) Gaussian blur, (2) Baseline JPEG compression, (3) Gaussian white noise, (4) JPEG2000 compression with dynamic contrast-based quantization, (5) Flat allocation (equal distortion contrast at all scales) and (6) Baseline JPEG2000 compression. Each distortion has been applied with three levels of intensity to generate distorted images.

A database MICT-Toyama presented in Ref. [15] has used 14 reference images to generate 168 distorted images. This database contains color images of a resolution of $768 \times 512$. Two distortions have been considered in this database including (1) JPEG compression and (2) JPEG2000 compression. Each distortion has been applied with six levels of intensity to generate distorted images. To get subjective ratings, a single stimulus methodology has been used.

A database IVC presented in Ref. [16] has used 10 reference images to generate 185 distorted images. This database contains color images of a resolution of $512 \times 512$. Four distortions have been considered in this database including (1) blurring, (2) local adaptive resolution (LAR) coding, (3) JPEG2000 compression and (4) JPEG compression. To get subjective ratings, a double stimulus methodology has been used.

A database TID2008 presented in Ref. [4] is an old version of TID2013 database. It contains 25 reference images to generate 1700 distorted images. This database contains color images of a resolution of $512 \times 384$. This database has used the same 17 distortions as TID2013 database. Also, the same methodology as TID2013 database has been used to get subjective ratings.

# References

[1]  S. Athar and Z. Wang, "A comprehensive performance evaluation of image quality assessment algorithms," *IEEE Access*, 7, 140030–140070, 2019.

[2]  H. Sheikh, M. Sabir, and A. Bovik, "A statistical evaluation of recent full reference image quality assessment algorithms," *IEEE Transactions on Image Processing*, 15(11), 3440–3451, 2006.

[3]  L. Cormack, H.R. Sheikh, Z. Wang, and A. Bovik, "LIVE image quality assessment database release 2", LIVE, University of Texas at Austin, 2006. Available: http://live.ece.utexas.edu/research/quality.

[4]  A. Zelensky, N. Ponomarenko, V. Lukin, K. Eziazarian, M. Carli, and F. Battisti, "TID2008– A database for evaluation of full reference image quality metrics," *Advance Modern Radioelectronics*, 10(4), 30–45, 2009.

[5]  E. C. Larson and D. M. Chandler, "Most apparent distortion: Full-reference image quality assessment and the role of strategy," *Journal of Electronic Imaging*, 19, 011006, 2010.

[6]  ITU-R Recommendation BT.500-13, "Methodology for the subjective assessment of the quality of television pictures," 2012.

[7]  N. Ponomarenko, L. Jin, O. Ieremeiev, V. Lukin, K. Egiazarian, J. Astola, B. Vozel, and Kacem, "Image database TID2013: Peculiarities, results and perspectives," *Signal Processing: Image Communication*, 30(C), 57–77, 2015.

[8]  A. Zaric, N. Tatalovic, N. Brajkovic, H. Hlevnjak, M. Loncaric, E. Dumic, and S. Grgic, "Vcl@fer image quality assessment database," in *IEEE International Conference on ELMAR-2011*, 2011, pp. 105–110.

[9]  X. Liu, M. Pedersen, and J. Y. Hardeberg, "CID:IQ – A new image quality database," in A. Elmoataz, O. Lezoray, F. Nouboud, and D. Mammass (Eds.) *Image and Signal Processing*, 2014, pp. 193–202.

[10]  D. Jayaraman, A. Mittal, A. K. Moorthy, and A. C. Bovik, "Objective quality assessment of multiply distorted images," in *IEEE International*

*Conference Record of the Forty Sixth Asilomar Conference on Signals, Systems and Computers (ASILOMAR)*, 2012, pp. 1693–1697.

[11]   K. Gu, G. Zhai, X. Yang, and W. Zhang, "Hybrid no-reference quality metric for singly and multiply distorted images," *IEEE Transactions on Broadcasting*, 60(3), 555–567, 2014.

[12]   W. Sun, F. Zhou, and Q. Liao, "MDID: A multiply distorted image database for image quality assessment," *Pattern Recognition*, 61, 153–168, 2017.

[13]   S. Corchs and F. Gasparini, "A multidistortion database for image quality," in S. Bianco, R. Schettini, A. Trémeau, and S. Tominaga (Eds.) *Computational Color Imaging*, 2017, pp. 95–104.

[14]   D. M. Chandler and S. S. Hemami, "VSNR: A wavelet-based visual signal-to-noise ratio for natural images," *IEEE Transactions on Image Processing*, 16(9), 2284–2298, 2007.

[15]   K. S. Y. Horita and K. Yoshikazu, "MICT image quality evaluation database."

[16]   P. L. Callet and F. Autrusseau, "Subjective quality assessment IRCCYN/IVC database," 2004.

# Chapter 5

# Subjective and Objective Image Quality Assessment

This chapter provides an overview of the subjective and objective image quality assessment techniques. It presents a classification of subjective and objective image quality assessment techniques. It starts with discussing the subjective image quality assessment techniques where it presents several international standards that are used in subjective image quality assessment along with a few methods which provide instruction and guidelines to measure subjective image quality. It further highlights the shortcomings of subjective image quality assessment techniques and introduces the need for objective quality assessment along with a few applications of it. Toward the end, this chapter introduces general purpose as well as distortion-specific image quality assessment mechanisms.

## 5.1 Subjective Image Quality Assessment

Subject image quality assessment involves human observers for quality assessment. It is considered the most reliable strategy for quality assessment of images as it involves human observers who are end users of most image and video applications. In the subjective way of testing and assessment of the quality of images, the quality score of each image is measured with the help of the opinions provided by a group of people. Figure 5.1 shows an example of subjective testing where a human observer

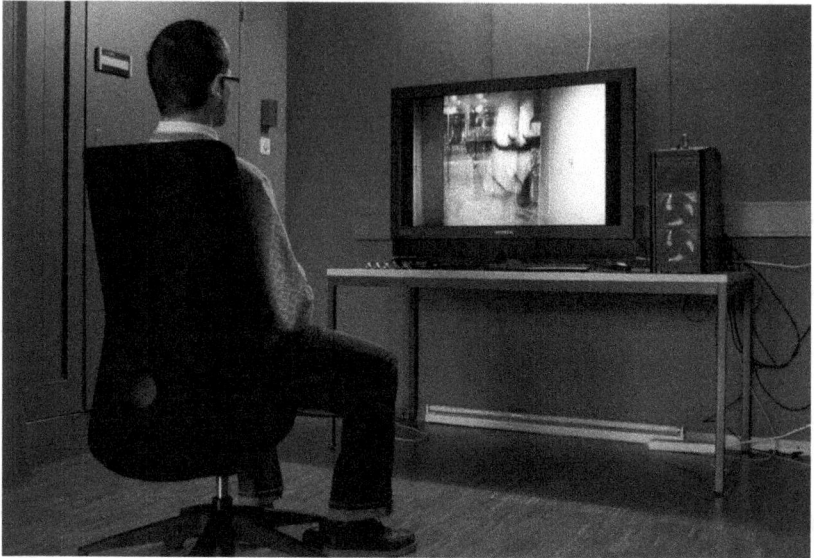

**Figure 5.1.**  An example of subjective testing: Quality test laboratory and 46″ polarized 3D display used for the subjective quality evaluation [1].

is deployed to look at the images on a computer screen and provide the quality of the image.

Several international standards such as [2–7] and [8] are available in the literature for reliable subjective testing. We discuss some of these standards in the following:

1. **ITU-R BT.500-11** [2]: This subjective quality assessment standard has been proposed for television pictures, and it is a widely used standard, which contains information about subjective experiments, presentation of subjective outcomes, test materials and viewing conditions.
2. **ITU-T P.910** [4]: This standard provides information about how to estimate quality for digital videos where the transmission rate is below 1.5 Mbits/sec.
3. **ITU-R BT.814-1** [5]: This standard helps in setting the contrast and brightness of display devices.
4. **ITU-R BT.1129-2** [6]: This standard provides instruction for quality assessment of standard definition (SD) video sequences.

There are a few methods which have been proposed in the literature to provide instruction and guidelines to measure subjective image quality.

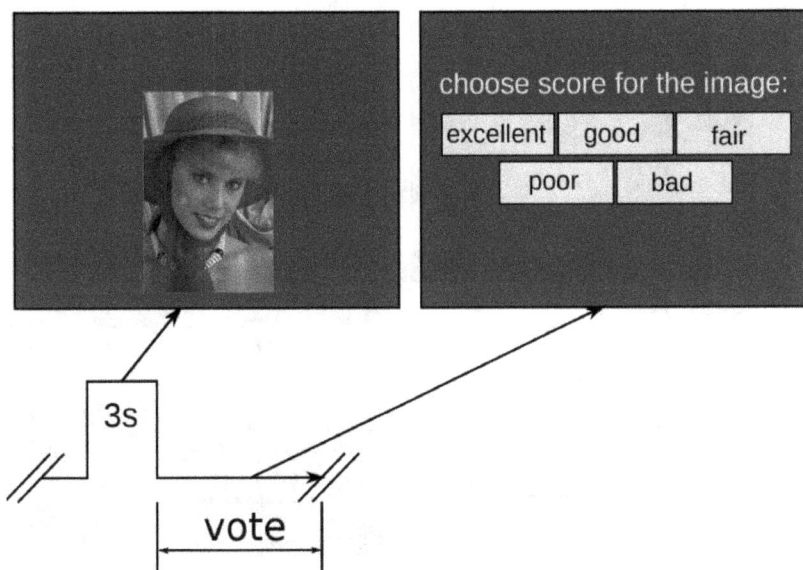

**Figure 5.2.**   An example of a single stimulus where the task is to provide quality of the displayed image [9].

We discuss some of these important subjective image quality assessment methods in the following subsections.

### 5.1.1   *Single stimulus categorical rating*

This method asks an observer to provide the quality for a displayed image that is presented for a fixed and short duration. An observer has to select one of the five categories, bad, poor, fair, good and excellent, for the displayed image. Figure 5.2 shows an example of a single stimulus rating where an image is displayed for 3 seconds, and then the user has to provide a quality score based on the observation. It is noted that all images, including a reference image, are displayed in random order and there is no time limit to vote for an image.

### 5.1.2   *Double stimulus categorical rating*

In this method, a reference image and a test image are displayed randomly one after another for a three-second duration. Figure 5.3 shows an example of a double stimulus rating where a good contrast image is displayed for

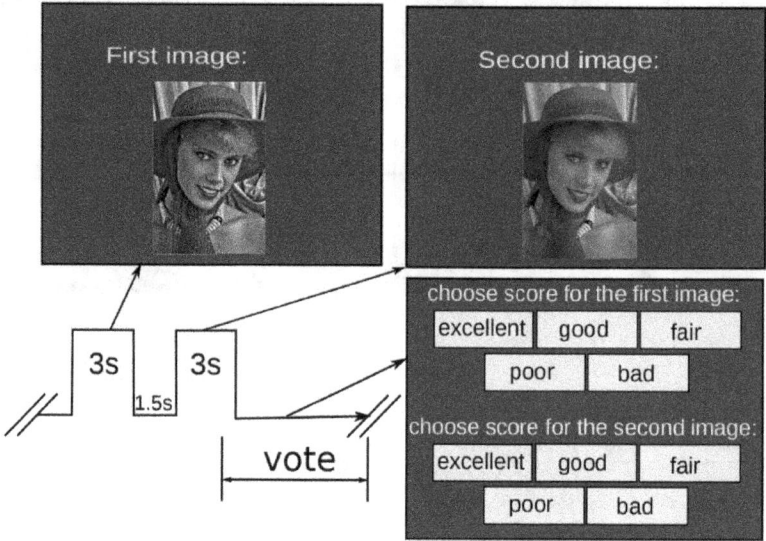

**Figure 5.3.** An example of a double stimulus where the task is to rate the quality of first and second image [9].

3 seconds. Then, after 1.5 seconds, a low-contrast version of the first image is displayed. Further, the observer was asked to vote for the quality of both images same as a single stimulus rating.

### 5.1.3 *Ordering by force-choice pair-wise comparison*

This method displayed a pair of images and asked the observer to select a higher-quality image. The observer always has to choose one image, even if both displayed images have the same quality (i.e., force-choice). Figure 5.4 shows two images simultaneously where the first image is a distorted image and the second is a good quality image. Now, an observer has to make the decision that which one has higher quality without any time limit. It is noted that there is no fixed time for displaying images; both images are displayed till the observer votes.

### 5.1.4 *Pair-wise similarity judgments*

The forced-choice method is used to select higher-quality images, but it cannot explain the difference between images. Pair-wise similarity judgments display a pair of images and ask an observer not only to select higher

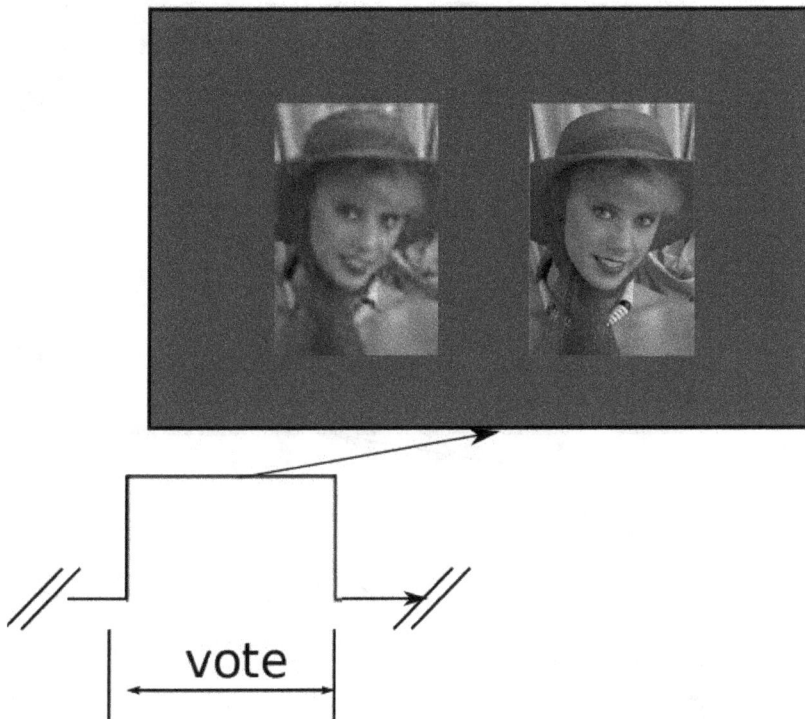

**Figure 5.4.** An example of force-choice where the task is to select a higher-quality image [9].

quality images but also to indicate how much difference in the quality. Figure 5.5 shows two images where the quality of first image is better than second image. The observer selects a higher-quality image and also marks the difference in quality on a continuous scale. In this figure, we can see that an observer has marked to "better" because the left image has better details and contrast compared to the right image.

It is worth noting that quality scores provided by subjective quality assessment techniques include subjectivity due to the involvement of human viewers in the task of image grading and quality assessment. Due to this, it has been pointed out by researchers [10] that these quality scores, if used directly, may be untrustworthy. This is due to several reasons. For example, human viewers may use different scaling for different types of images. Further, it is possible that the scale may vary for different types of images distortion. To handle these problems, it is suggested to use some kind of normalization to bring the scores of different subjects to a common

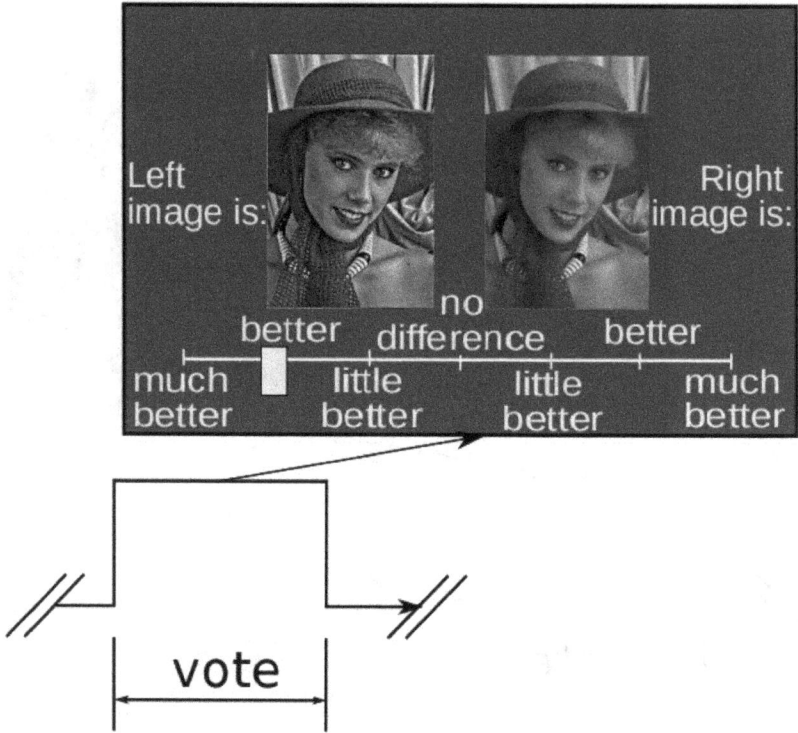

**Figure 5.5.** An example of pair-wise similarity where the task is to assess the quality difference between the two images [9].

scale. Two techniques, namely Difference Mean Opinion Score (DMOS) and Z-score, are used in the literature for this purpose and have been discussed in Chapter 1 under Section 1.4 in detail.

### 5.1.5   *Limitations of subjective image quality assessment*

All the above-mentioned methods provide reliable and accurate subjective assessments. However, subjective estimation has several limitations:

1. Many observers are involved in subjective testing; therefore, it is expensive and time-consuming. This makes subjective image quality assessment impractical in real-world applications.
2. Subjective testing cannot be incorporated as a pre-processing step in any image-based applications, such as image compression.

3. Human mood can change the result of subjective testing. Moreover, illumination conditions and display devices may influence the outcome.
4. Subjective image quality assessment further becomes tricky due to the involvement of several factors, such as the type of device used for display, environmental conditions, vision ability of the subjects, mood of the subjects and viewing distance.

Hence, it is always advisable to design and use mathematical models which can imitate the image quality assessment capability of an average human observer and automatically estimate the perceptual quality of an image consistent with subjective evaluations. The objective image quality assessment techniques are developed in the same line and offer image quality assessment based on mathematical models that are capable of predicting image quality automatically and accurately. It is expected from an ideal objective image quality assessment technique to do quality predictions at par with that offered by an average human observer.

## 5.2 Objective Image Quality Assessment

The objective quality assessment estimates perceptual quality of an image automatically using a mathematical model. For an objective image quality assessment technique to be an ideal image quality assessment technique, it must be able to mimic the quality predictions of an average human observer. Objective image quality assessment techniques can be classified into three categories, *viz.* full-reference image quality assessment (FR-image quality assessment), reduced-reference image quality assessment (RR-image quality assessment) and no-reference image quality assessment (NR-image quality assessment). This division is based on the requirement of the availability of the reference image in the quality estimation process. Wherever used, the reference image is believed to be of perfect quality with no distortions present in it.

Figure 5.6 shows the flow of each objective image quality assessment technique. As we can see, the quality of an image in NR-image quality assessment techniques is estimated without using any reference image. This shows that there is no requirement for any reference image in NR-image quality assessment techniques. In contrast to this, FR-image quality assessment techniques consider the availability of a reference image to

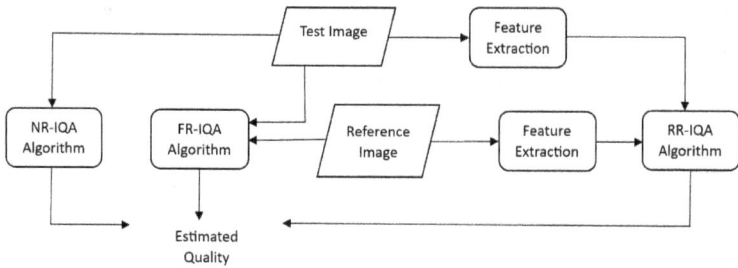

**Figure 5.6.** Objective image quality assessment approaches [11].

assess the quality of a given image. The RR-image quality assessment techniques take the middle ground where they consider that reference image is not fully available. Alternatively, the availability of some features extracted from the reference image is considered and is used for quality estimation of an image. Hence, the RR-image quality assessment techniques rely on the features of reference as well as the given test image to estimate its quality. Since in real-world scenarios, the reference image is mostly not available, NR-image quality assessment techniques are the most preferred choice in practice. Chapters 6, 7 and 8 discuss FR-image quality assessment, RR-image quality assessment and NR-image quality assessment techniques, respectively, in detail.

Image quality assessment techniques can also be classified based on the scope of their application. This way, we can have two categories, *viz.*, general-purpose image quality assessment techniques and distortion-specific image quality assessment techniques. The general-purpose image quality assessment techniques perform quality estimation of images in the presence of any kind of distortion. On the other hand, distortion-specific image quality assessment techniques estimate quality of an image for a particular distortion.

### 5.2.1   *Applications of objective image quality assessment*

Objective image quality assessment techniques are widely used as an automatic quality assessment in many applications [13]:

1. Quality control system needs to check quality of images before using them further in any image-based application. For example, objective image quality assessment can be used in image acquisition systems to obtain the best quality image using estimated quality. Figure 5.7 shows

(a) $Q_I = 0.2328$

(b) $Q_I = 0.4348$

(c) $Q_I = 0.7312$

(d) $Q_I = 0.6287$

**Figure 5.7.** Few examples of images with bad illumination (degraded due to dark spots) with their estimated quality score $Q_I$ [12].

some poor-quality images affected by bad illumination. A method proposed in Ref. [12] has estimated quality based on illumination, which helps select or reject captured images before their use.

2. Objective image quality assessment methods can be used to select an algorithm out of many enhancement algorithms by selecting higher-quality images from their output.

3. Objective image quality assessment methods may be used to optimize transmission and image processing systems. In a visual communication network, at the encoder, we can make use of image quality assessment metrics to optimize pre-filtering and bit assignment algorithms. At the decoder, image quality assessment metric is used to optimize post-filtering and reconstruction algorithms.

### 5.2.2   *Challenges of objective image quality assessment*

While numerous image quality assessment (IQA) metrics have demonstrated strong performance on existing databases, it's important to recognize that the image quality measurement problem remains complex. We contend that current subjective databases may not address all applications, particularly in light of emerging challenges for image quality measurement:

1. Assessing the quality of an inpainted image involves various image editing tasks, such as filling in image impairments with adjacent texture or inferred structure, flawlessly removing objects from the scene and seamlessly fusing two image patches. Designing a metric for evaluating the quality of inpainted images is challenging. While the metric can draw insights from distortion criteria inherent in the problem formulation, such as the continuity of gradient, optical flow field, edges, contours and structures, there are still additional features that can be leveraged. Advancements in quality metrics are undoubtedly beneficial for enhancing inpainting algorithms.

2. State-of-the-art image coding algorithms have integrated several advanced techniques, including spatio-temporal Just Noticeable Difference (JND) models, object-oriented coding, shape/texture coding and fovea coding. There is a growing demand for a new rate-distortion theory to inform the optimization of coding schemes. Visual attention plays a pivotal role in quality assessment, as regions of interest receive the majority of bits during coding. Texture similarity is also essential for

the metric, as traditional methods struggle to evaluate the quality of textured regions. Implementing intelligent coding techniques for texture may lead to further bit savings.

3. Image perceptual watermarking is a fascinating area of study that leverages various properties of the Human Visual System (HVS) to embed more payload of invisible watermarks. Watermarks are strategically designed as flexible and intelligent noise, often concentrated in very low or high-frequency sub-bands, distributed across bright or dark regions, and exhibiting non-uniform characteristics in other features sensitive to the HVS. In applications involving covert communication, a looser distortion constraint can be tolerated. The covert message is embedded as a watermark, aiming to evade detection by eavesdroppers by avoiding semantic distortions or abnormal flaws. In such scenarios, the watermarked image can be deemed acceptable even if its traditional quality remains low. A metric that thoroughly exploits the properties of the HVS holds the potential to revolutionize watermarking and covert communication.

4. Image adaptation is a critical area of research that enables universal multimedia access (UMA) on communication networks. During the adaptation process, changes in image resolution and aspect ratio may occur, but it is imperative to ensure that the main foreground remains undistorted and is displayed with an appropriate size. Consequently, the adapted image may experience non-uniform geometric distortion. If the quality degradation resulting from image adaptation can be accurately assessed, it could serve as a valuable similarity metric for scene retrieval.

# References

[1]  L. Goldmann, F. De Simone, and T. Ebrahimi, "Impact of acquisition distortion on the quality of stereoscopic images," 2010.

[2]  ITU-R Recommendation BT.500-11,"Methodology for the subjective assessment of the quality of television pictures", ITU, Geneva, Switzerland, 2002.

[3]  ITU-R Recommendation BT.710-4, "Subjective assessment methods for image quality in high-definition television," ITU, Geneva, Switzerland, 1998.

[4]  ITU-T Recommendation P.910, "Subjective video quality assessment methods for multimedia applications," ITU, Geneva, Switzerland, 2008.

[5] ITU-R Recommendation BT.814-1, "Specification and alignment procedures for setting of brightness and contrast of displays," ITU, 1994.

[6] ITU-R Recommendation BT.1129-2, "Subjective assessment of standard definition digital television (SDTV) systems," ITU, 1998.

[7] ITU-R Recommendation BT.1361, "Worldwide unified colorimetry and related characteristics of future television and imaging systems," ITU, 1998.

[8] ITU-R Recommendation BT.815-1, "Specification of a signal for measurement of the contrast ratio of displays," ITU, 1994.

[9] R. K. Mantiuk, A. Tomaszewska, and R. Mantiuk, "Comparison of four subjective methods for image quality assessment," *Computer Graphics Forum*, 31(8), 2478–2491, 2012.

[10] A. M. van Dijk, J.-B. Martens, and A. B. Watson, "Quality assessment of coded images using numerical category scaling," in *Advanced Image and Video Communications and Storage Technologies*, 2451, International Society for Optics and Photonics. SPIE, 1995, pp. 90–101.

[11] G. Ciocca, S. Corchs, F. Gasparini, and R. Schettini, "How to assess image quality within a workflow chain: an overview," *International Journal on Digital Libraries*, 15, 1–25, 2014.

[12] P. Joshi and S. Prakash, "Image enhancement with naturalness preservation," *Visual Computer*, 36(1), 71–83, 2020.

[13] Z. Wang and A. C. Bovik, *Modern Image Quality Assessment*. Synthesis Lectures on Image, Video & Multimedia Processing, Morgan & Claypool Publishers, California, 2006.

# Chapter 6

# Full Reference Image Quality Assessment

Full reference-based image quality assessment metrics compare the input test image against a pristine reference image where presumably no distortion is present. These techniques estimate the quality of a distorted image by computing the difference between an original image (reference image) and a distorted image. The literature on FR-image quality assessment is classified into five categories:

1. **Error visibility methods**: The main idea behind these methods is to measure the distance between pixels or transformed representations of distorted and reference images to estimate the perceptual quality of a distorted image. This category's most widely used method is a mean square error (MSE). MSE exhibits certain characteristics that contribute to its widespread use as a performance measure in signal processing. Following are some of these attributes:

    (a) It is a straightforward and computationally efficient approach.
    (b) It provides a clear physical interpretation, representing the energy of an error signal in a natural manner.
    (c) Due to its fulfilment of properties such as convexity, symmetry and differentiability, MSE is highly regarded as a quality measure in optimization tasks.
    (d) It is widely accepted as a standard, commonly utilized for optimization and evaluation across various signal processing applications.

Despite the above interesting features, MSE is poorly correlated with perceptual quality. This is attributed to the omission of certain significant physiological and psychophysical traits of the human visual system (HVS) in this measure. This can be elucidated from Figure 6.1, which shows MSE values for differently distorted images. It is noted that the MSE values of several of the distorted images are nearly identical even though the same images present dramatically (and obviously) different visual quality [1]. Implicit assumptions are inherent when employing the MSE measure, rendering it inadequate for assessing image quality. These assumptions are outlined as follows:

(a) If the reference and test images undergo a random reordering in a comparable manner, the MSE between them will remain unaltered. This highlights that MSE is unaffected by the temporal or spatial arrangement of samples within the reference image.
(b) For a given distortion signal, the MSE remains constant irrespective of the reference signal to which it is added.
(c) MSE is unaffected by the sign of the error signal samples.
(d) When computing MSE, image signals are treated with equal importance.

To overcome MSE limitations, many image quality assessment techniques (such as [2–7] and [8]) have been developed that first map images to another representation and then compute MSE in that representation.

2. **Structural similarity (SSIM) methods**: This category of methods measures the similarity between a local region of distorted and a reference image. The well-known standard image quality assessment method is SSIM [9], which considers three significant components: an image's structure, contrast and luminance. Over the years, researchers have modified and extended SSIM to better estimate perceptual image quality. For example, feature similarity [10], gradient similarity [11], edge strength similarity [12] and saliency similarity [13] are the advanced extension of SSIM. Figure 6.2 shows SSIM values for differently distorted images. SSIM value varies from 0 to 1, where 1 indicates perfect similarity and 0 indicates no similarity. Figure 6.2 (b) results from little contrast stretched; therefore, its SSIM with the original image is high. On the other hand, Figure 6.2 (e) is a highly blurred image; therefore, its SSIM value is low, which depicts less similarity of this image with the original image.

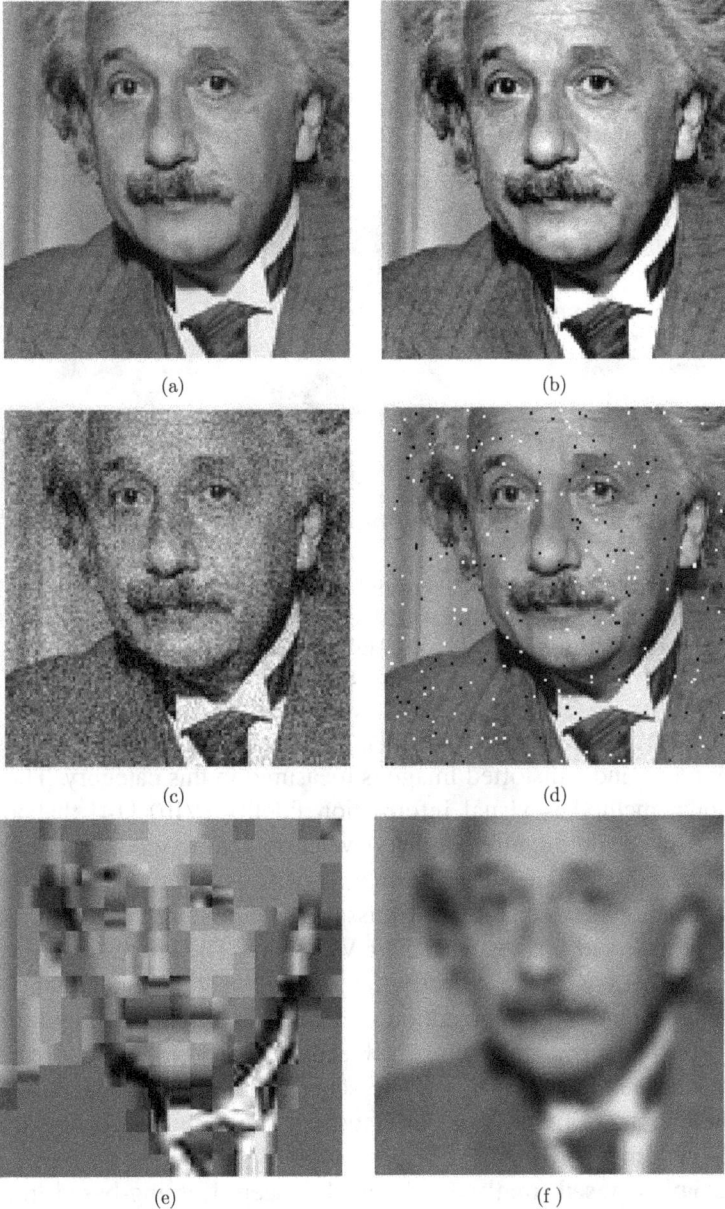

**Figure 6.1.** MSE values for differently distorted images [1]: (a) Original image; (b) contrast-stretched image, MSE = 306; (c) Gaussian noise contamination, MSE = 309; (d) impulsive noise contamination, MSE = 313; (e) JPEG compression, MSE = 309; (f) blurred image, MSE = 308.

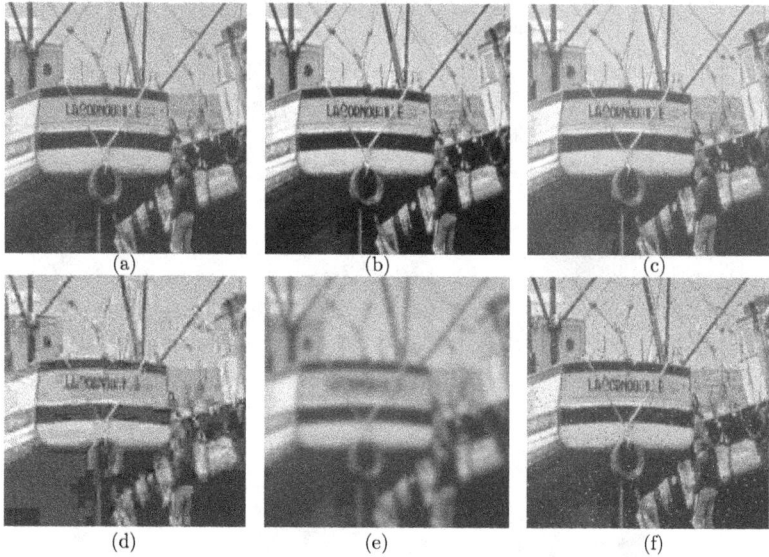

**Figure 6.2.** SSIM values for differently distorted images [9]: (a) Original image; (b) contrast-stretched image, SSIM = 0.9168; (c) mean-shifted image, SSIM = 0.9900; (d) JPEG compressed image, SSIM = 0.6949; (e) blurred image, SSIM = 0.7052; (f) salt-pepper impulsive noise-contaminated image, SSIM = 0.7748.

3. **Information-theoretic methods**: The mutual information between a reference and a distorted image is measured in this category. The well-known method is visual information fidelity (VIF) [14] that applied Gaussian scale mixtures in the wavelet domain to model the reference and the distorted images. Further, quality is estimated using mutual information between the two Gaussian scale mixtures. Figure 6.3 shows a few distorted images with their VIF score. The interesting feature of VIF is that it can also quantify the improvement in visual quality, such as an estimation of degradation in quality. For example, Figure 6.3 (a) is an original image having a VIF score of 1 and Figure 6.3 (b) shows its enhanced image having a VIF score of 1.1 after contrast stretching. The increment in the VIF score above 1 shows an improvement in the original image.

4. **Learning-based methods**: Recently, deep learning-based methods have been developed for image quality estimation. These methods learn from a training set of images and corresponding perceptual quality scores. For example, a technique presented in [15] has extracted spatial and frequency domain features from pairs of images (reference and

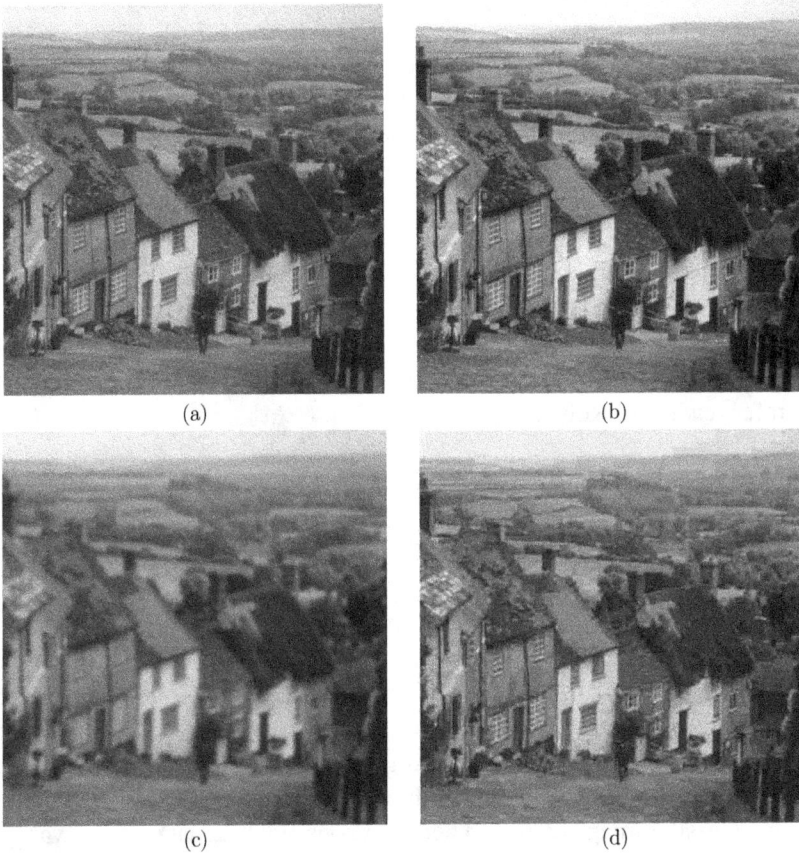

**Figure 6.3.** VIF values for differently distorted images [14]: (a) Original image VIF = 1; (b) contrast-stretched image, VIF = 1.10; (c) blurred image, VIF = 0.07; (d) JPEG compression, VIF = 0.10.

distorted) and fused them by Random Forest (RF) to estimate quality. Convolutional neural network (CNN)-based image quality assessment method is proposed in Ref. [16]. This method extracts deep features from patches of reference and distorted images and fuses them to assess the image's quality. By leveraging the power of deep neural networks, these methods have achieved state-of-the-art performance on existing image-quality databases.

5. **Fusion-based methods**: In this category, existing FR-image quality assessment methods are fused and developed into a new FR-image quality assessment estimator similar to boosting in machine learning. For

example, a technique proposed in Ref. [17] has studied R-SVD, MS-SSIM and VIF FR-image quality assessment methods and presented fusion by arithmetic expression. Regression-based fusion technique has been utilized in Ref. [18] for the effective FR-image quality assessment. The multiple linear regression-based fusion and kernel ridge regression-based fusion have been proposed in Refs. [19] and [20], respectively. In technique [21], six FR-image quality assessment techniques have been fused using a trained neural network. Genetic algorithm-based fusion and multi-gene genetic programming-based fusion have been employed in Refs. [22] and [23], respectively.

We have given a glimpse of FR-image quality assessment methods, for comprehensive surveys about FR-image quality assessment, we refer readers to Refs. [24], [25] and [26].

## 6.1   Full-Reference Image Quality Assessment of Color Images

The objective Full-Reference Image Quality Assessment (FR-IQA) methods discussed thus far are specifically tailored for grayscale images and do not utilize color information inherent in images. Color information plays a crucial role in simplifying the identification and extraction of objects within a scene, thereby influencing human observers' judgments when assessing image quality. There is a consistent demand across various domains dealing with digital images for objective quality metrics capable of predicting the quality of a color image in comparison to its reference version. Such metrics find applications in computer graphics for comparing the photorealism levels of different rendering methods, image coding for evaluating the performance of diverse compression schemes, image processing to assess color image enhancement methods and false-color multi-spectral image fusion [27].

Although objective IQA metrics designed for grayscale images can, in theory, be extended to encompass color images, this is typically achieved by applying these metrics individually to each of the three RGB color channels and then combining the quality scores for each channel. However, this straightforward approach does not align with human perception, primarily because the RGB color space does not accurately represent color as perceived by the Human Visual System (HVS) [27].

The initial color image quality measure is introduced in Ref. [28]. This research presents a basic model of human color vision that quantitatively explains various perceptual parameters, such as brightness and saturation. The perceptual space is conceptualized as a vector space with spatial filtering attributes. Additionally, a norm is introduced on this vector space, facilitating the measurement of distances and defining a distortion metric that aligns closely with perceptual assessments. Further studies focusing on color image quality assessment can be found in Refs. [29], [30] and [31].

The FSIM index is intended for use with grayscale images or the luminance component of color images. To adapt the FSIM index for color images, the reference RGB color image is first converted into another color space where the luminance component can be isolated from the chrominance. In Ref. [10], the RGB color image is transformed into the YIQ color space, where Y represents the luminance component and I and Q represent the chrominance components. Further, similarity between chrominance components of a reference image and a test image is measured to estimated the quality of a color image.

## 6.2 Full-Reference Image Quality Assessment of High Dynamic Range (HDR) Images

In recent years, there has been a growing interest in High Dynamic Range (HDR) images, which offer a wider range of intensity values compared to Low Dynamic Range (LDR) images. To visualize HDR images on standard display devices, tone mapping operators (TMOs) [32] [33] [34] are utilized. However, as TMOs reduce the dynamic range of HDR images, they often lead to information loss and quality degradation. Hence, it is crucial to evaluate the quality of each tone-mapped image to determine which TMO yields superior-quality LDR images. Moreover, with the emergence of various display technologies such as HDR displays, digital cinema projections and mobile devices' displays, there is a need to assess the quality of images with different dynamic ranges. This evaluation helps in gauging the capability of each display device to produce high-quality images.

Subjective evaluation stands out as the most dependable approach for gauging the quality of HDR and LDR images [35] [36] [37]. Nonetheless, as previously highlighted, these methods are costly, time-intensive and impractical for integration into optimization algorithms. Thus, there is a pressing need to devise objective Image Quality Assessment (IQA) techniques to evaluate the quality of HDR images and their associated tone-mapped renditions. The full-reference IQA methods discussed earlier cannot fulfill this objective because they presuppose similarity in dynamic range between the reference and test images.

Two commonly employed full-reference image quality assessment methods are utilized for assessing the quality of images possessing varying dynamic ranges. These methods include the dynamic range independent quality measure (DRIM) [38], tailored for assessing the quality of images across arbitrary dynamic ranges, and the tone-mapped images quality index (TMQI) [39], specifically designed to evaluate the quality of tone-mapped images in relation to their reference HDR images.

In this section, we assess the predictive capabilities of the Full-Reference Image Quality Assessment (FR-IQA) methods discussed in preceding sections: PSNR, SSIM [8], MS-SSIM [9], VIF [10], MAD [11], FSIM [12], FSIMC [12] and TMQI [14]. The original MATLAB implementations of these methods, as provided by their respective authors, have been utilized for evaluation purposes.

Tables 6.1–6.3 present the test findings of the six Full-Reference Image Quality Assessment (FR-IQA) methods across three subjective quality datasets for an overall evaluation of the performance of the considered image quality metrics. Tables 6.4 and 6.5 provide the average Spearman Rank-Order Correlation Coefficient (SRCC), Kendall's Rank Correlation Coefficient (KRCC), Pearson Linear Correlation Coefficient (PLCC), Root Mean Square Error (RMSE) and Mean Absolute Error (MAE) results over three datasets. These average values are computed in two scenarios. In the first scenario, the scores of the performance measures are directly averaged.

**Table 6.1.** Performance evaluation of FR-IQA algorithms on CSIQ dataset.

| | CSIQ Dataset | | | | |
| --- | --- | --- | --- | --- | --- |
| | KRCC | SRCC | PLCC | MAE | RMSE |
| SSIM | 0.6907 | 0.8756 | 0.8613 | 0.0991 | 0.1334 |
| PSNR | 0.6084 | 0.8058 | 0.8000 | 0.1195 | 0.1575 |
| MAD | 0.7970 | 0.9466 | 0.9502 | 0.0636 | 0.0818 |
| FSIM | 0.7567 | 0.9242 | 0.9120 | 0.0797 | 0.1077 |
| VIF | 0.7537 | 0.9195 | 0.9277 | 0.0743 | 0.0980 |
| MS-SSIM | 0.7393 | 0.9133 | 0.8991 | 0.0870 | 0.1149 |

**Table 6.2.** Performance evaluation of FR-IQA algorithms on LIVE dataset.

| | LIVE Dataset | | | | |
|---|---|---|---|---|---|
| | KRCC | SRCC | PLCC | MAE | RMSE |
| SSIM | 0.7963 | 0.9479 | 0.9449 | 6.9325 | 8.9455 |
| PSNR | 0.6865 | 0.8756 | 0.8723 | 10.5093 | 13.3597 |
| MAD | 0.8421 | 0.9669 | 0.9675 | 5.2202 | 6.9037 |
| FSIM | 0.8337 | 0.9634 | 0.9597 | 5.9468 | 7.6780 |
| VIF | 0.8282 | 0.9636 | 0.9604 | 6.1070 | 7.6137 |
| MS-SSIM | 0.8045 | 0.9513 | 0.9489 | 6.6978 | 8.6188 |

**Table 6.3.** Performance evaluation of FR-IQA algorithms on TID 2008 dataset.

| | TID 2008 Dataset | | | | |
|---|---|---|---|---|---|
| | KRCC | SRCC | PLCC | MAE | RMSE |
| SSIM | 0.5768 | 0.7749 | 0.7732 | 0.6547 | 0.8511 |
| PSNR | 0.4027 | 0.5531 | 0.5734 | 0.8327 | 1.0994 |
| MAD | 0.6445 | 0.8340 | 0.8308 | 0.5562 | 0.7468 |
| FSIM | 0.6946 | 0.8805 | 0.8738 | 0.4926 | 0.6525 |
| VIF | 0.5860 | 0.7491 | 0.8084 | 0.6000 | 0.7899 |
| MS-SSIM | 0.6568 | 0.8542 | 0.8451 | 0.5578 | 0.7173 |

**Table 6.4.** Average performance over three datasets.

| | Direct Average | | | | |
|---|---|---|---|---|---|
| | KRCC | SRCC | PLCC | MAE | RMSE |
| SSIM | 0.6879 | 0.8661 | 0.8598 | 2.5621 | 3.3100 |
| PSNR | 0.5659 | 0.7448 | 0.7486 | 3.8205 | 4.8722 |
| MAD | 0.7612 | 0.9158 | 0.9162 | 1.9467 | 2.5774 |
| FSIM | 0.7617 | 0.9227 | 0.9152 | 2.1730 | 2.8127 |
| VIF | 0.7226 | 0.8774 | 0.8988 | 2.2604 | 2.8339 |
| MS-SSIM | 0.7335 | 0.9063 | 0.8977 | 2.4475 | 3.1503 |

In the second scenario, different weights are assigned to various datasets based on their sizes, measured in terms of the number of images (1700 for TID2008, 866 for CSIQ and 779 for LIVE datasets, respectively).

**Table 6.5.**    Average performance over three datasets.

| | Dataset Size-Weighted Average | | | | |
|---|---|---|---|---|---|
| | KRCC | SRCC | PLCC | MAE | RMSE |
| SSIM | 0.6574 | 0.8413 | 0.8360 | 1.9729 | 2.5504 |
| PSNR | 0.5220 | 0.6943 | 0.7017 | 2.9016 | 3.7018 |
| MAD | 0.7300 | 0.8941 | 0.8935 | 1.5148 | 2.0085 |
| FSIM | 0.7431 | 0.9111 | 0.9037 | 1.6559 | 2.1476 |
| VIF | 0.6858 | 0.8432 | 0.8747 | 1.7464 | 2.1999 |
| MS-SSIM | 0.7126 | 0.8921 | 0.8833 | 1.8658 | 2.4015 |

**Table 6.6.**    Average performance over three datasets.

Computation Time for an image of size $512 \times 512$ (in seconds/image)

| | SSIM | PSNR | MAD | FSIM | VIF | MS-SSIM |
|---|---|---|---|---|---|---|
| Time | 0.0293 | 0.0035 | 2.0630 | 0.3508 | 1.3647 | 0.0834 |

We also assessed the computation time for each chosen FR-IQA method. The average time required to evaluate the quality of images sized $512 \times 512$ pixels was measured. Experiments were conducted on a laptop equipped with an Intel Core i7 processor running at 1.6 GHz, using the MATLAB R2013a software platform. The results are detailed in Table 6.6.

## 6.3   Screening of Full-Reference Image Quality Assessment Models

In this section, we briefly evaluate existing top FR-image quality assessment techniques used for reference image recovery. Using an initial image $y_0$ and undistorted image $x$, techniques aimed to recover $x$ by numerically optimization:

$$y^* = arg_y \ min \ D(x, y) \tag{6.1}$$

where $y^*$ is the recovered image and $D$ is defined as the FR-image quality assessment technique where a lower score indicates higher estimated quality. Suppose, our FR-image quality assessment is MSE and analytical solution is $y^* = x$, then it indicates the full recoverability. As most of the image quality assessment algorithms are continuous and differentiable, a gradient-based iterative method is used to obtain the solution numerically. We present seventeen FR-image quality assessment methods used for the recovery of the reference image, including five DNN techniques DISTS [40], LPIPS [41], PieAPP [42], DeepIQA [16] and GTI-CNN [43], two information-theoretical techniques VIF [14] and IFC [44], seven structural similarity techniques MCSD [45], VSI [13], GMSD [46], SFF [47], FSIM [10], CW-SSIM [48] and MS-SSIM [49], and three error visibility techniques NLPD [50], PAMSE [51] and MAD [52].

Figures 6.4 and 6.5 present the results of reference image recovery for 17 FR-image quality assessment methods. Figure 6.4 considers a white Gaussian noise image and Figure 6.5 considers a JPEG-compressed image of a reference image as an initialization. In all FR-image quality assessment methods, optimization converges when the recovered image has a substantially better quality score than the initial distorted image. Only a few injective mapping-based FR-image quality assessment techniques, including MS-SSIM, PAMSE, NLPD and DISTS, are able to recover the

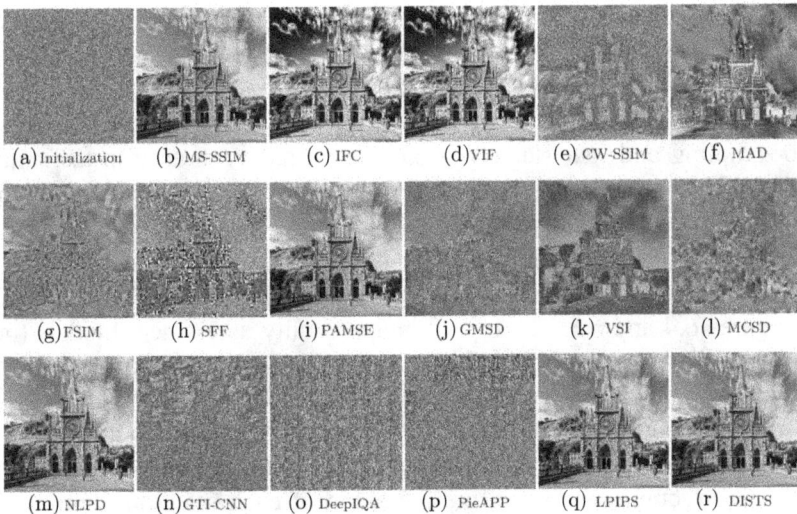

**Figure 6.4.** Recovery of a reference image [53]. An image (a) is a Gaussian noise image. Images from (b) to (r) are recovered images obtained using different image quality assessment algorithms by optimizing estimated quality relative to a reference image.

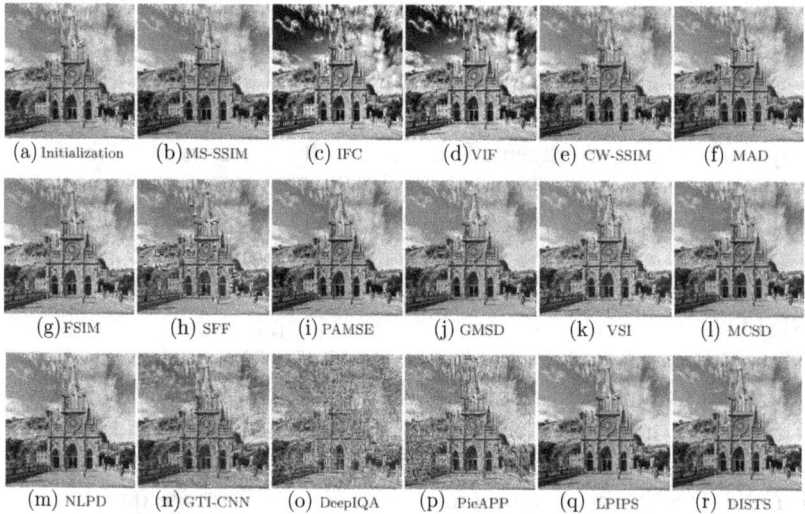

**Figure 6.5.**   Recovery of a reference image [53]. An image (a) JPEG compressed version of a reference image. Images from (b) to (r) are recovered images obtained using different image quality assessment algorithms by optimizing estimated quality relative to a reference image.

reference image. Many remaining techniques have failed to recover the image and even generated poorer quality than initial image (see Figures 6.5 (o) and (p)). It is because FR-image quality assessment depends on surjective mapping functions that represent an image on reduced "perceptual" space [53]. For example, surjective DNN with half-wave rectification, sub-sampling and four stages of convolution has been used in technique GTI-CNN [54] that generated results at the cost of significant information loss. Other techniques, such as PieAPP [42] and DeepIQA [16], have also used surjective DNN-based image quality assessment models.

We have shown results for 17 FR-image quality assessment techniques in Figures 6.4 and 6.5. The top 10 image quality assessment models (out of 17) that have generated good results are discussed in the following:

1. MS-SSIM is the multi-scale extension of SSIM method allowing a wider range of viewing distances compared to a single scale. Contrast and structure similarities are measured at each scale and luminance similarity is measured at the coarsest scale from the Gaussian pyramids (i.e., decomposed version of an input image). MS-SSIM has proved its robustness in estimating perceptual quality of the image. Therefore, it

has been used in many applications such as compression algorithms [55] and designing of DNN-based image super-resolution [56].

2. VIF is used to quantitatively measure the information preserved in the distorted image from the reference image. Natural image statistics that are represented using a Gaussian scale mixture and mutual information is measured between the distorted image and the reference image with an assumption of additive noise perturbations and signal attenuation. VIF is a robust technique that can handle a case even when the quality of the distorted image is better than the reference image [57].

3. CW-SSIM method is used to estimate the quality when an image is distorted by geometric distortions, such as rotations and translation. To preserve image features, this method consists in shifting a local phase of wavelet coefficients.

4. An human visual system-based method called MAD (most apparent distortion) has used contrast and luminance masking to measure near-threshold distortions. In addition, spatial-frequency statistics are employed to measure supra-threshold distortions. Further, both measurements are fused using a weighted geometric mean where weight indicates the amount of distortion.

5. FSIM (Feature SIMilarity) method considers low-level features of an image to estimate quality. Two types of features are extracted including phase congruency as the primary feature and gradient magnitude as the complementary feature. FSIM provides its color version image quality assessment using chromatic components.

6. GMSD (gradient magnitude similarity deviation) simply computes gradient magnitude of pixels in an image and then performed standard deviation pooling. The advantage of GMSD is the low-cost quality computation. However, the pooling can be failed when distortions make a standard deviation of zero.

7. VSI (visual saliency induced) method focuses on salient regions for estimating quality. It assumes that visual quality is proportional to change in salient regions of the image due to presence of distortions. This method prepares a salient map as a weighting function to define the significance of a local region. VSI combines chromatic features, gradient magnitude and saliency magnitude to estimate the quality score for the image.

8. NLPD (normalized Laplacian pyramid distance) is an HVS-based method that transforms local gain control and local luminance subtraction nonlinearly. Further, this method fuses these transformations

using weighted p-norms. NLPD method has been used to optimize a compression system [58] and image rendering algorithms [59] for higher dynamic range input image.

9. LPIPS (learned perceptual image patch similarity) method first represents a distorted and a reference image into deep spaces using deep neural network (DNN) and then computes Euclidean distance between them. This paper showed that features computed by DNN are able to mimic human perception of image quality.

10. DISTS (deep image structure and texture similarity) is based on an injective mapping function that fuses texture similarity measurements and SSIM-like structure of a reference image and a distorted image. This image quality assessment method is robust to modest geometric transformations and texture resampling. However, it is sensitive to structural distortions.

## 6.4 Quality Assessment of 3D Images

The availability of digital 3D images for consumer consumption has seen rapid growth in recent years. According to statistics from the Motion Picture Association of America (MPAA), in 2011, half of all moviegoers watched at least one 3D movie, with those under 25 years old viewing more than twice that number. To meet this escalating demand, the production of 3D movies has been increasing by at least 50 percent annually in recent years. Beyond movies, other forms of 3D content are becoming increasingly prevalent in our daily lives through 3D television broadcasts and 3D on mobile devices. However, these contents introduce a range of complex technological and perceptual challenges. Achieving consistent, comfortable and believable depth perception requires determining a multitude of parameters in the imaging and processing stages in a perceptually meaningful manner. Nonetheless, due to inevitable trade-offs in real-world applications, the visual quality of these 3D contents may degrade. Therefore, to uphold and enhance the Quality of Experience (QoE) of 3D visual contents, both subjective and objective quality assessment methods are imperative. These methods are of significant importance for display manufacturers, content providers and service providers alike. Compared to its 2D counterpart, 3D Image Quality Assessment (IQA) faces a plethora of new challenges, including depth perception, virtual view synthesis and asymmetric stereo compression.

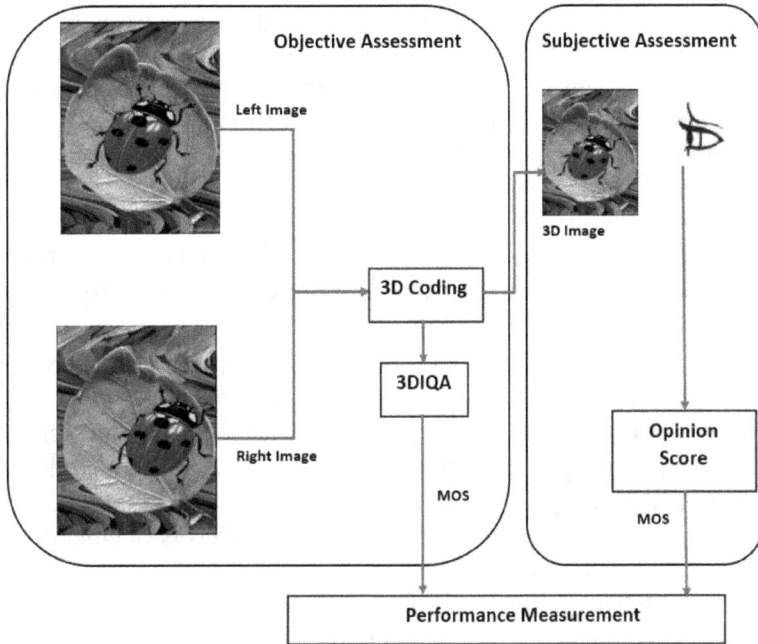

**Figure 6.6.** Methodology for Stereoscopic Image Quality Assessment.

An important query arises regarding the suitability of 2D Image Quality Assessment (IQA) methods for evaluating 3D images. Investigations conducted in Refs. [60] and [61] aim to address this inquiry. The findings indicate that 2D objective IQA techniques are effective in assessing the quality of 3D images only when dealing with symmetric images, where the Peak Signal-to-Noise Ratios (PSNRs) of the two-eye images are nearly identical. Figure 6.6 shows the methodology to assess the quality of stereoscopic images, which is divided into three parts:

1. **Subjective Assessment**: Observers will provide their opinions on the quality of the stereoscopic images (the input). This assessment is subjective because different observers may evaluate the same stereoscopic pair differently based on their individual visual perception.
2. **Objective Assessment of a 3D Coding Process using 3D Image Quality Assessment (3DIQA)**: This assessment is considered objective, as the same stereoscopic pair will yield consistent evaluation results regardless of the observer.

3. **Correlation between Objective and Subjective Assessments**: The strength of the relationship between these two assessments will be measured to estimate how closely the 3DIQA results align with human observers' opinions. The higher the correlation, the more accurately the objective evaluation reflects the average human perception of image quality.

Some of the proposed quality descriptors of 3-D contents that quantify the overall viewing experience of a 3-D representation are as follows [62]:

1. **Depth quality**: It is essential to scrutinize the depth characteristics of 3D data to ensure the content's suitability for viewing [63].
2. **Naturalness**: This refers to the threshold where viewers can seamlessly merge left and right views into a lifelike 3D image with a fluid depth representation [64].
3. **Presence**: A realistic 3D scene heightens the viewers' feeling of being present within the environment.
4. **Value-add**: The perceived advantage of presenting content in 3D compared to displaying the same content in 2D [65].
5. **Discomfort**: The overall subjective perception arising from physiological and/or psychological effects of viewing 3D content [65].
6. **Overall 3D Quality of Experience (QoE)**: Typically measured in terms of Differential Mean Opinion Score (DMOS).

It's worth mentioning that there are currently no widely accepted methods for quantifying the aforementioned descriptors. Nonetheless, efforts have been made recently to tackle this issue through the introduction of standards. In the following, we outline some of these standards:

1. The ITU-R [66] has issued a new recommendation regarding the subjective quality assessment of 3-D TV systems. This recommendation primarily focuses on picture quality, depth quality and visual comfort.
2. The VQEG is actively involved in three primary areas: establishing ground truth data for validating subjective evaluation methodologies, verifying objective 3-D video quality evaluation and assessing the impact of viewing environments on 3-D quality evaluation.
3. IEEE has initiated efforts toward developing a standard for assessing the quality of 3-D content, displays and devices based on human factors.

This endeavor examines various characteristics of displays, devices, environments, content and viewers.

The classification of 2-D IQA methods, namely FR-IQA, RR-IQA and NR-IQA, can be applied to 3-D images, albeit with some differences in their definitions [67]. This stems from the inherent difficulty in accessing the reference and test 3-D images as they are perceived. As we can only access the left and right views of a scene and not the cyclopean image, which is a single mental image generated by the brain through combining the images received from both eyes, this presents a challenge for both reference and test cyclopean images. Consequently, the problem of 3-D IQA is characterized as double-blind [67].

The initial objective IQA for 3-D images is introduced in Ref. [68]. This method employs reliable 2-D IQA techniques, including SSIM, UQI, methods described in Ref. [69], and the metric outlined in Ref. [70]. However, it's important to note that this approach doesn't consider the depth information of 3-D images. Based on the utilized information, 3-D IQA methods can be categorized into two groups: those relying solely on color information and those incorporating both color and disparity information [71].

### 6.4.1  *Methods based on color information*

The techniques falling within this category rely solely on color information [72–76]. In Ref. [72], quality scores are computed based on the SIFT-matched feature points. A multiple channel model is employed to estimate 3-D image quality in [73]. Technique in Ref. [74] presents an RR-IQA method for 3-D images that utilizes extracted edge information. Modeling the Gabor response of binocular vision is the focus for measuring 3-D image quality [75]. Lastly, a state-of-the-art 3-D IQA method for video compression is proposed in Ref. [76].

### 6.4.2  *Methods based on color and disparity information*

The techniques in this group utilize both color and disparity information to assess the overall quality of 3-D data [77–79]. Reference [77] introduces an RR-IQA method for 3-D images based on eigenvalue or eigenvector analysis. Additionally, Refs. [78] and [79] propose two NR-IQA methods for 3-D images.

### 6.4.3 *Subjective 3-D image quality datasets*

In this subsection, several subjective 3-D image quality datasets are introduced. The LIVE 3-D IQA dataset [80] comprises 20 reference images, 5 distortion categories and a total of 365 test images. Quality scores in this dataset are in the form of DMOS. It is the first publicly available 3-D IQA dataset incorporating true-depth information, stereoscopic pairs and human opinion scores. Distortion types include JPEG compression, JPEG2000 compression, additive white Gaussian noise, Gaussian blur and a fast-fading model based on the Rayleigh fading channel.

The IVC 3-D image dataset [81] contains 6 reference images and 15 distorted versions of each image along with their respective subjective scores. Distortion types include JPEG compression, JPEG2000 compression and blurring. The dataset comprises a total of 96 images.

The only known dataset for HDR 3-D images and their corresponding tone-mapped versions is available in Ref. [82]. This dataset includes 9 reference 3-D HDR images tone-mapped using 8 TMOs, resulting in a total of 81 images. Additionally, the dataset provides statistics of these images (including min, max and mean luminance) and their histograms.

# References

[1]  Z. Wang and A. C. Bovik, "Mean squared error: Love it or leave it? a new look at signal fidelity measures," *IEEE Signal Processing Magazine*, 26(1), 98–117, 2009.

[2]  R. Safranek and J. Johnston, "A perceptually tuned sub-band image coder with image dependent quantization and post-quantization data compression," in *International Conference on Acoustics, Speech, and Signal Processing,* vol. 3, 1989, pp. 1945–1948.

[3]  S. J. Daly, "Visible differences predictor: An algorithm for the assessment of image fidelity," in B. E. Rogowitz (Ed.), *Human Vision, Visual Processing, and Digital Display III*, International Society for Optics and Photonics, 1992, pp. 2–15.

[4]  J. Lubin, *The Use of Psychophysical Data and Models in the Analysis of Display System Performance*. Cambridge, MA, USA: MIT Press, 1993, pp. 163–178.

[5] A. B. Watson, "Dctune: A technique for visual optimization of dct quantization matrices for individual images." 1993.

[6] P. C. Teo and D. J. Heeger, "Perceptual image distortion," in B. E. Rogowitz and J. P. Allebach, (Eds.) *Human Vision, Visual Processing, and Digital Display V*, International Society for Optics and Photonics, 1994, pp. 127–141.

[7] A. Watson, G. Yang, J. Solomon, and J. Villasenor, "Visibility of wavelet quantization noise," *IEEE Transactions on Image Processing*, 6(8), 1164–1175, 1997.

[8] E. C. Larson and D. M. Chandler, "Most apparent distortion: full-reference image quality assessment and the role of strategy," *Journal of Electronic Imaging*, 19, 011006, 2010.

[9] Z. Wang, A. Bovik, H. Sheikh, and E. Simoncelli, "Image quality assessment: From error visibility to structural similarity," *IEEE Transactions on Image Processing*, 13(4), 600–612, 2004.

[10] L. Zhang, L. Zhang, X. Mou, and D. Zhang, "FSIM: A feature similarity index for image quality assessment," *IEEE Transactions on Image Processing*, 20(8), 2378–2386, 2011.

[11] D. Glasner, S. Bagon, and M. Irani, "Super-resolution from a single image," in *IEEE 12th International Conference on Computer Vision*, 2009, pp. 349–356.

[12] X. Zhang, X. Feng, W. Wang, and W. Xue, "Edge strength similarity for image quality assessment," *IEEE Signal Processing Letters*, 20(4), 319–322, 2013.

[13] L. Zhang, Y. Shen, and H. Li, "VSI: A visual saliency-induced index for perceptual image quality assessment," *IEEE Transactions on Image Processing*, 23(10), 4270–4281, 2014.

[14] H. Sheikh and A. Bovik, "Image information and visual quality," *IEEE Transactions on Image Processing*, 15(2), 430–444, 2006.

[15] Z. Tang, Y. Zheng, K. Gu, K. Liao, W. Wang, and M. Yu, "Full-reference image quality assessment by combining features in spatial and frequency domains," *IEEE Transactions on Broadcasting*, 65(1), 138–151, 2019.

[16] S. Bosse, D. Maniry, K.-R. Müller, T. Wiegand, and W. Samek, "Deep neural networks for no-reference and full-reference image quality assessment," *IEEE Transactions on Image Processing*, 27(1), 206–219, 2018.

[17] K. Okarma, "Combined full-reference image quality metric linearly correlated with subjective assessment," in *Artificial Intelligence and Soft Computing*, 2010, pp. 539–546.

[18] K. Okarma, "Extended hybrid image similarity – combined full-reference image quality metric linearly correlated with subjective scores," *Elektronika Ir Elektrotechnika*, 19, 129–132, 2013.

[19]   M. Oszust, "Image quality assessment with lasso regression and pairwise score differences," *Multimedia Tools and Applications*, 76, 13 255–13 270, 2016.

[20]   Y. Yuan, Q. Guo, and X. Lu, "Image quality assessment: A sparse learning way," *Neurocomputing*, 159, 227–241, 2015.

[21]   V. V. Lukin, N. N. Ponomarenko, O. I. Ieremeiev, K. O. Egiazarian, and J. Astola, "Combining full-reference image visual quality metrics by neural network," in *Human Vision and Electronic Imaging*, vol. 9394, International Society for Optics and Photonics. SPIE, 2015, pp. 172–183.

[22]   M. Oszust, "Decision fusion for image quality assessment using an optimization approach," *IEEE Signal Processing Letters*, 23(1), 65–69, 2016.

[23]   N. Merzougui and L. Djerou, "Multi-gene genetic programming based predictive models for full-reference image quality assessment," *Journal of Imaging Science and Technology*, vol. 65, pp. 060409-1–060409-13, 2021.

[24]   M. Pedersen and J. Hardeberg, "Full-reference image quality metrics: Classification and evaluation," *Foundations and Trends® in Computer Graphics and Vision*, 7, 1–80, 01 2012.

[25]   L. Zhang, L. Zhang, X. Mou, and D. Zhang, "A comprehensive evaluation of full reference image quality assessment algorithms," in *2012 19th IEEE International Conference on Image Processing*, 2012, pp. 1477–1480.

[26]   V. Wasson and B. Kaur, "Full reference image quality assessment from iqa datasets: A review," in *2019 6th International Conference on Computing for Sustainable Global Development (INDIACom)*, 2019, pp. 735–738.

[27]   A. Toet and M. P. Lucassen, "A new universal colour image fidelity metric," *Displays*, 24(4), 197–207, 2003.

[28]   O. Faugeras, "Digital color image processing within the framework of a human visual model," *IEEE Transactions on Acoustics, Speech, and Signal Processing*, 27(4), 380–393, 1979.

[29]   P. L. Callet and D. Barba, "Perceptual color image quality metric using adequate error pooling for coding scheme evaluation," In B. E. Rogowitz and T. N. Pappas, (Eds.) *Human Vision and Electronic Imaging VII*, vol. 4662, International Society for Optics and Photonics. SPIE, 2002, pp. 173–180.

[30]   Y.-K. Lai, J. Guo, and C.-C. J. Kuo, "Perceptual fidelity measure of digital color images," in B. E. Rogowitz and T. N. Pappas, (Eds.) *Human Vision and Electronic Imaging III*, vol. 3299, International Society for Optics and Photonics. SPIE, 1998, pp. 221–231.

[31]   M.-S. Lian, "Image evaluation using a color visual difference predictor (CVDP)," *Proc SPIE*, 4299, 175–186, 06 2001.

[32]   E. Reinhard, M. Stark, P. Shirley, and J. Ferwerda, "Photographic tone reproduction for digital images," *ACM Transactions on Graphics*, 21(3), 267–276, 2002.

[33] G. Larson, H. Rushmeier, and C. Piatko, "A visibility matching tone repro-duction operator for high dynamic range scenes," *IEEE Transactions on Visualization and Computer Graphics*, 3(4), 291–306, 1997.

[34] R. Fattal, D. Lischinski, and M. Werman, "Gradient domain high dynamic range compression," *ACM Transactions on Graphics*, 21(3), 249–256, 2002.

[35] F. Drago, W. L. Martens, K. Myszkowski, and H.-P. Seidel, "Perceptual evaluation of tone mapping operators," in *ACM SIGGRAPH 2003 Sketches & Applications*, 2003, p. 1.

[36] J. Kuang, H. Yamaguchi, C. Liu, G. M. Johnson, and M. D. Fairchild, "Evaluating HDR rendering algorithms," *ACM Transactions on Applied Perception*, 4(2), 2007.

[37] P. Ledda, A. Chalmers, T. Troscianko, and H. Seetzen, "Evaluation of tone mapping operators using a high dynamic range display," *ACM Transactions on Graphics*, 24(3), 640–648, 2005.

[38] T. O. Aydin, R. Mantiuk, K. Myszkowski, and H.-P. Seidel, "Dynamic range independent image quality assessment," *ACM Transactions on Graphics*, 27(3), 1–10, 2008.

[39] H. Yeganeh and Z. Wang, "Objective quality assessment of tone-mapped images," *IEEE Transactions on Image Processing*, 22(2), 657–667, 2013.

[40] K. Ding, K. Ma, S. Wang, and E. P. Simoncelli, "Image quality assessment: Unifying structure and texture similarity," *IEEE Transactions on Pattern Analysis and Machine Intelligence*, 44, 2567–2581, 2020.

[41] R. Zhang, P. Isola, A. Efros, E. Shechtman, and O. Wang, "The unreason-able effectiveness of deep features as a perceptual metric," 01 2018.

[42] E. Prashnani, H. Cai, Y. Mostofi, and P. Sen, "Pieapp: Perceptual image-error assessment through pairwise preference," *IEEE/CVF Conference on Computer Vision and Pattern Recognition*, pp. 1808–1817, 2018.

[43] K. Ma, Z. Duanmu, and Z. Wang, "Geometric transformation invariant image quality assessment using convolutional neural networks," in *IEEE International Conference on Acoustics, Speech and Signal Processing (ICASSP)*, 2018, pp. 6732–6736.

[44] H. Sheikh, A. Bovik, and G. de Veciana, "An information fidelity criterion for image quality assessment using natural scene statistics," *IEEE Transactions on Image Processing*, 14(12), 2117–2128, 2005.

[45] T. Wang, L. Zhang, H. Jia, B. Li, and H. Shu, "Multiscale contrast similar-ity deviation: An effective and efficient index for perceptual image quality assessment," *Signal Processing: Image Communication*, 45, 1–9, 2016.

[46] W. Xue, L. Zhang, X. Mou, and A. C. Bovik, "Gradient magnitude simi-larity deviation: A highly efficient perceptual image quality index," *IEEE Transactions on Image Processing*, 23(2), 684–695, 2014.

[47] H.-W. Chang, H. Yang, Y. Gan, and M.-H. Wang, "Sparse feature fidelity for perceptual image quality assessment," *IEEE Transactions on Image Processing*, 22(10), 4007–4018, 2013.

[48] Z. Wang and E. Simoncelli, "Translation insensitive image similarity in complex wavelet domain," in *Proceedings IEEE International Conference on Acoustics, Speech, and Signal Processing*, 2, 573–576, 2005.

[49] Z. Wang, E. Simoncelli, and A. Bovik, "Multiscale structural similarity for image quality assessment," in *Proceedings of the Thrity-Seventh Asilomar Conference on Signals, Systems and Computers*, 2, 1398–1402, 2003.

[50] V. Laparra, J. Ballé, A. Berardino, and E. P. Simoncelli, "Perceptual image quality assessment using a normalized Laplacian pyramid," in *HVEI*, 2016.

[51] W. Xue, X. Mou, L. Zhang, and X. Feng, "Perceptual fidelity aware mean squared error," in *IEEE International Conference on Computer Vision*, 2013, pp. 705–712.

[52] E. Larson and D. Chandler, "Most apparent distortion: Full-reference image quality assessment and the role of strategy," *Journal of Electronic Imaging*, 19, 011006, 01 2010.

[53] K. Ding, K. Ma, S. Wang, and E. P. Simoncelli, "Comparison of full-reference image quality models for optimization of image processing systems," 129(4), 1258–1281, 2021.

[54] K. Ma, Z. Duanmu, and Z. Wang, "Geometric transformation invariant image quality assessment using convolutional neural networks," in *2018 IEEE International Conference on Acoustics, Speech and Signal Processing (ICASSP)*, 2018, pp. 6732–6736.

[55] J. Ballé, D. C. Minnen, S. Singh, S. J. Hwang, and N. Johnston, "Variational image compression with a scale hyperprior," *ArXiv*, vol. abs/1802.01436, 2018.

[56] H. Zhao, O. Gallo, I. Frosio, and J. Kautz, "Loss functions for image restoration with neural networks," *IEEE Transactions on Computational Imaging*, 3(1), 47–57, 2017.

[57] S. Wang, K. Ma, H. Yeganeh, Z. Wang, and W. Lin, "A patch-structure representation method for quality assessment of contrast changed images," *IEEE Signal Processing Letters*, 22(12), pp. 2387–2390, 2015.

[58] J. Ballé, V. Laparra, and E. P. Simoncelli, "End-to-end optimization of nonlinear transform codes for perceptual quality," in *2016 Picture Coding Symposium (PCS)*, 2016, pp. 1–5.

[59] K. Ma, H. Yeganeh, K. Zeng, and Z. Wang, "High dynamic range image compression by optimizing tone mapped image quality index," *IEEE Transactions on Image Processing*, 24(10), 3086–3097, 2015.

[60] C. Hewage, "Prediction of stereoscopic video quality using objective quality models of 2-d video," *Electronics Letters*, 44(2), 963–965, 2008.

[61] C. T. E. R. Hewage, S. T. Worrall, S. Dogan, S. Villette, and A. M. Kondoz, "Quality evaluation of color plus depth map-based stereoscopic video," *IEEE Journal of Selected Topics in Signal Processing*, 3(2), 304–318, 2009.

[62] S. Winkler and D. Min, "Stereo/multiview picture quality: Overview and recent advances," *Signal Processing: Image Communication*, 28(10), 1358–1373, 2013.

[63] P. Lebreton, A. Raake, M. Barkowsky, and P. Le Callet, "Evaluating depth perception of 3d stereoscopic videos," *IEEE Journal of Selected Topics in Signal Processing*, 6(6), 710–720, 2012.

[64] P. J. Seuntiëns, I. E. J. Heynderickx, W. A. IJsselsteijn, P. M. J. van den Avoort, J. Berentsen, I. J. Dalm, M. T. Lambooij, and W. Oosting, "Viewing experience and naturalness of 3D images," in *Three-Dimensional TV, Video, and Display IV*, vol. 6016. SPIE, 2005, p. 601605.

[65] M. Lambooij, W. Ijsselsteijn, and I. Heynderickx, "Visual discomfort in stereoscopic displays: A review," *Proceedings of SPIE, 2007 vol. 6490*, 53, 2007.

[66] J. Zhou, G. Jiang, X. Mao, M. Yu, F. Shao, Z. Peng, and Y. Zhang, "Subjective quality analyses of stereoscopic images in 3dtv system," *International Telecommunication Union*, 11 2011.

[67] A. K. Moorthy, C.-C. Su, A. Mittal, and A. C. Bovik, "Subjective evaluation of stereoscopic image quality," *Signal Processing: Image Communication*, 28(8), 870–883, 2013.

[68] A. Benoit, P. Le Callet, P. Campisi, and R. ..., "Quality assessment of stereoscopic images," *Eurasip Journal on Image and Video Processing - EURASIP J Image Video Process*, vol. 2008, 12 2008.

[69] M. Carnec, P. Le Callet, and D. Barba, "An image quality assessment method based on perception of structural information," in *Proceedings of International Conference on Image Processing*, vol. 3, III–185, 2003.

[70] Z. Wang and E. Simoncelli, "Reduce-reference image quality assessment using a wavelet-domain natural image statistic model," *Proceedings of SPIE - The International Society for Optical Engineering*, vol. 5666, 03 2005.

[71] Y.-H. Lin and J.-L. Wu, "Quality assessment of stereoscopic 3d image compression by binocular integration behaviors," *IEEE Transactions on Image Processing*, 23(4), 1527–1542, 2014.

[72] P. Gorley and N. Holliman, "Stereoscopic image quality metrics and compression," In J. Woods, N. S. Holliman, and J. O. Merritt, (Eds.) *Stereoscopic Displays and Applications XIX*, A., vol. 6803, International Society for Optics and Photonics. SPIE, 2008, p. 680305.

[73] L. Shen, J. Yang, and Z. Zhang, "Stereo picture quality estimation based on a multiple channel hvs model," in *2009 2nd International Congress on Image and Signal Processing*, 2009, pp. 1–4.

[74] C. T. E. R. Hewage and M. G. Martini, "Reduced-reference quality metric for 3d depth map transmission," in *2010 3DTV-Conference: The True Vision - Capture, Transmission and Display of 3D Video*, 2010, pp. 1–4.

[75] R. Bensalma and C. Larabi, "A perceptual metric for stereoscopic image quality assessment based on the binocular energy," *Multidimensional Systems and Signal Processing*, vol. 24, 2013.

[76] V. De Silva, H. K. Arachchi, E. Ekmekcioglu, and A. Kondoz, "Toward an impairment metric for stereoscopic video: A full-reference video quality metric to assess compressed stereoscopic video," *IEEE Transactions on Image Processing*, 22(9), 3392–3404, 2013.

[77] A. Maalouf and M.-C. Larabi, "Cyclop: A stereo color image quality assessment metric," in *IEEE International Conference on Acoustics, Speech and Signal Processing (ICASSP)*, 2011, pp. 1161–1164.

[78] J. Si, B. Huang, H. Yang, W. Lin, and Z. Pan, "A no-reference stereoscopic image quality assessment network based on binocular interaction and fusion mechanisms," *IEEE Transactions on Image Processing*, 31, 3066–3080, 2022.

[79] M.-J. Chen, L. K. Cormack, and A. C. Bovik, "No-reference quality assessment of natural stereopairs," *IEEE Transactions on Image Processing*, 22(9), 3379–3391, 2013.

[80] C.-C. S. A. K. Moorthy and A. C. Bovik, "Live 3d image quality dataset." Available: http://live.ece.utexas.edu/research/quality/live_3dimage.html, 2013.

[81] P. C. A. Benoit, P. Le Callet and R. Cousseau, "IVC 3d images dataset." Available: https://ivc.uwaterloo.ca/database/3DIQA.html.

[82] Z. Mai, C. Doutre, P. Nasiopoulos, and R. K. Ward, "Rendering 3-d high dynamic range images: Subjective evaluation of tone-mapping methods and preferred 3-d image attributes," *IEEE Journal of Selected Topics in Signal Processing*, 6(5), 597–610, 2012.

# Chapter 7

# Reduced-Reference Image Quality Assessment

Reduced-reference image quality assessment methods need some features (i.e., edges and textures) of the reference image to estimate the quality of a distorted image. Over the last decade, RR-based image quality assessment techniques have become very popular in various applications, such as video streaming, online games and social media. In RR-image quality assessment, it is not compulsory to have access to the reference image for the estimation of quality. Instead, features based on characteristics of pixels, dominant features or some coefficients of certain transformations (such as wavelets) are required. RR-image quality assessment methods can be considered the intermediate approach between FR-image quality assessment and NR-image quality assessment methods as it requires partial information to estimate quality. In today's world, the end-user demands high-quality multimedia content. Therefore, RR-image quality assessment methods are useful in estimating the degradation of image quality at the receiver end [1] [2]. For this, with the transmission of multimedia content, its certain extracted features are also sent over a medium. At the receiver's end, similar features (such as the sender's end) are extracted, and the RR-image quality assessment method is employed to assess quality between features sent from the sender's side and features extracted at the receiver's side. The estimated quality quantifies the degradation in the perceptual quality at the receiver's end compared to the sender's. A large number of features must be sent to the receiver's side (computationally high) to measure quality with high accuracy. On the other hand, fewer features take less time to communicate but may affect the performance of an RR-image quality assessment method. An example of a requirement for high-quality multimedia content is Internet service providers. These service providers must agree on providing a standard level of quality content. In this scenario, RR-image quality assessment methods can play a significant role in monitoring live streaming systems [3] [4]. In addition, many standard compression algorithms have been developed to compress multimedia content. RR-image quality assessment methods are useful for assessing quality of multimedia content compressed by these compression algorithms. Reduced-Reference Image Quality Assessment (RR-IQA) is valuable in applications where only partial reference data are available, and real-time or efficient quality monitoring is crucial. Here are some detailed applications:

1. **Real-Time Video Streaming**: In live streaming, it's critical to monitor image quality without slowing down transmission. RR-IQA uses small reference data portions, minimizing bandwidth while identifying quality drops due to compression, resolution changes or network issues.
2. **Wireless Communication Networks**: Wireless channels are prone to interference, compression artifacts and signal degradation. RR-IQA allows quality monitoring with reduced data, ensuring signals are adjusted promptly to maintain clarity, essential for mobile networks, satellite communications and 5G applications.
3. **Telemedicine and Remote Diagnostics**: Medical images sent across networks can lose quality, impacting diagnostics. RR-IQA helps monitor and verify image quality, ensuring that essential visual information is preserved, even with limited data, to maintain diagnostic accuracy.
4. **Content Delivery Networks (CDNs)**: For adaptive streaming services, RR-IQA evaluates video and image quality under fluctuating network conditions. By assessing using partial data, RR-IQA aids in adjusting streaming quality in real time, balancing clarity and bandwidth efficiency.
5. **Smart Surveillance Systems**: In security applications, RR-IQA monitors image quality to detect issues such as blurring or noise in video feeds, enabling quick adjustments in surveillance systems and ensuring that critical details in footage remain visible.

This chapter provides recent developments in RR-based quality estimation methods and is useful for researchers working in multimedia quality monitoring. In the following subsection, we present the classification of RR-image quality assessment methods.

# 7.1 Classification of RR Quality Assessment Methods

RR-image quality assessment methods have used different features based on different scenarios, which is useful to classify these methods into meaningful classes. RR-image quality assessment methods have mainly used pixels, frequency, etc. operations to assess image quality. Pixel-based image quality assessment methods use one or more pixels at a time to estimate quality, while frequency-based methods represent the original image into frequency space for quality assessment. Pixel-based image quality assessment methods are simple compared to frequency-based image quality assessment methods. However, in some scenarios, frequency-based methods provide better results than pixel-based methods because frequency space reveals more characteristics of an image. We have included three meaningful categories of RR-image quality assessment methods: pixel-based, frequency-based and 3D multimedia-based image quality assessment, as shown in Figure 7.1. Later, a pixel-based method is divided into point-based and mask-based methods. Similarly, frequency-based methods are classified into discrete wavelet and discrete

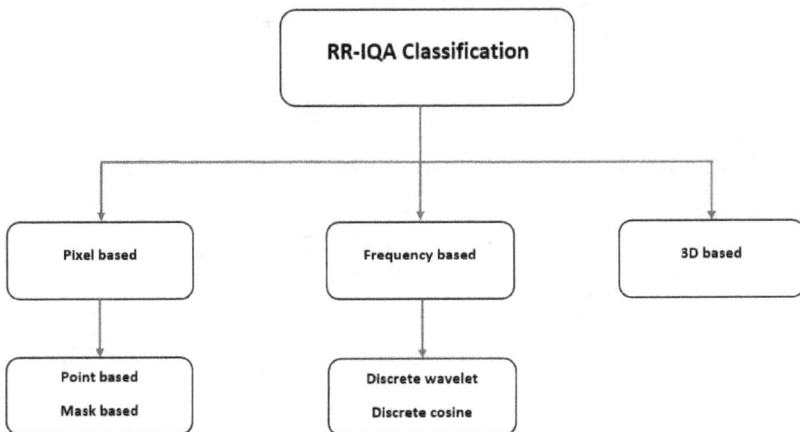

**Figure 7.1.** Classification of RR-image quality assessment.

cosine methods. Three-dimensional area for image quality assessment is less explored, so we present 3D in a single class. The classification is discussed in the following subsections.

## 7.1.1    *Pixel-based image quality assessment methods*

The literature contains many RR-image quality assessment techniques that manipulate pixels of an image to assess quality. Point-based methods focus on a pixel, whereas mask-based methods consider more than one pixel, i.e., considering neighboring pixels of a pixel. These methods are easy to implement, and their complexity is proportional to the number of pixels they consider. Mean square error (MSE) [5] and peak signal-to-noise ratio (PSNR) [6] are two widely used pixel-based methods. They are from the beginning of the image processing field and also used in quality assessment.

### 7.1.1.1    *Point-based methods*

Point-based image quality assessment methods focus on a particular pixel and change its value. A technique proposed in Ref. [7] has extracted features such as visual information and disorder certainty from original and distorted images to estimate the quality. The framework of this method is shown in Figure 7.2. A technique presented in Ref. [8] has

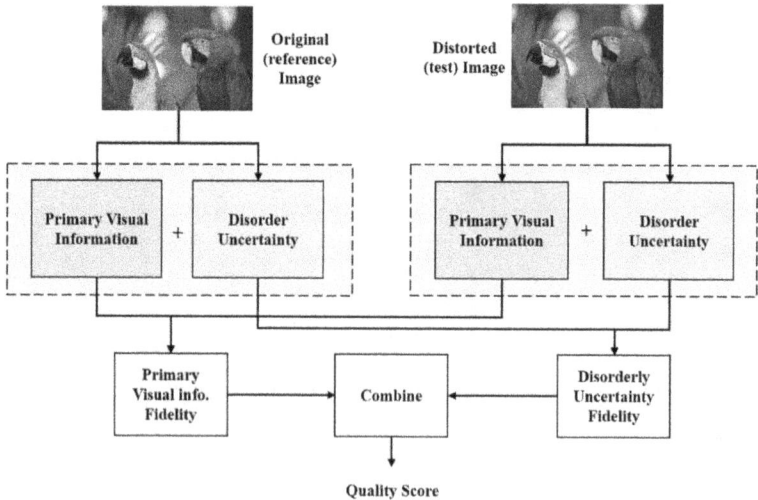

**Figure 7.2.**    Pixel-based RR-image quality assessment methods with respect to an original (reference) and a distorted (test) image [7].

utilized orientation selectivity (OS)-based features as HVS-based features. In OS-based features, this technique analyzes the similarity of orientations between two nearby pixels and then an OS-based visual pattern is formed to represent visual information. Further, a histogram is computed using visual information for a reference image and a distorted image. The difference between these histograms exhibits the quality score. The higher difference indicates the presence of higher distortions. Phase congruency variation between a reference and a distorted image is employed to measure image's quality. This method first computes fractal dimensions [9] on phase congruency of input images as features, and then these features are characterized by spatial distribution to estimate quality score. Techniques proposed in Refs. [2] and [10] have measured loss in temporal information and spatial information between a reference image and a distorted image. A technique named energy variation descriptor (EVD) is proposed in Ref. [11] to estimate spatial perspective by quantifying the change in energy in each video frame. For a temporal perspective, the generalized Gaussian density (GGD) function is used to obtain natural statistics of the inter-frame histogram. Further, city-block distance (CBD) measures the distances between the original video sequence and the encoded one. Finally, spatial EVD is combined with temporal CBD to estimate the quality score.

### 7.1.1.2  *Mask-based methods*

Different portions of an image are considered for computation in mask-based operations. Features such as contrast, width, length, local mean value and linear structure orientation are computed for different sub-images. The final output (estimated quality) is the contributions of operations on different portions. It has been seen that the structure information of an image is very sensitive to image distortions [12]. So, structure information can play a significant role in assessing image quality. Techniques proposed in Refs. [12] and [13] have used perceptual structural information for RR-based quality estimation of an image and a video. An RR-image quality assessment-based technique presented in Ref. [14] has transformed an image into a multi-scale orientation and then extracted statistical features from them to assess quality. A technique proposed in Ref. [15] has used the relation between quality of the light field and depth map for assessing the distorted image's quality. Another technique proposed in Ref. [5] has computed image statistics based on Weibull distribution and gradient magnitude of an image to assess quality. Singular value decomposition (SVD) has been used to compute the luminance information of an image. SVD can also be used to extract features from the image, and these features help in estimating quality of a distorted image [16]. A JPEG and JPEG2000 compression-based estimation technique has been proposed in Ref. [17] that can compute low data-rate features of a reference image. This technique has used fast Johnson—Lindenstrauss transform (FJLT)-based image hashing and is categorized into three main steps: (i) random sampling, (ii) dimension reduction and (iii) weights incorporation. In the first step, if the input is color image, then it is converted to a grayscale image and N sub-images are assigned in the image using a secret key. These sub-images were further mapped into lower dimensions using FJIT and assigned weights randomly to hash features. The final information is the indication of quality of the input image. A two-layer-based RR-image quality assessment technique proposed in Ref. [18] has used color correlogram to analyze distortion in color images. The first layer's task is to detect the type of distortion, whereas the second layer predicts the quality score for the identified distortion. The framework of the two-layer-based RR-image quality assessment technique is presented in Figure 7.3. A Video Quality Model (VQM) is proposed in Ref. [19] that has computed harmonic information using harmonic analysis of edges in an image. A receiver receives this harmonic information as a feature of a video. Further, using the received video, the receiver computed the same harmonic information and compared it with receiver's side harmonic information to estimate degradation in quality. Some RR-image quality assessment techniques such as those in Refs. [20] and [21] have used mask operations for feature extraction. Also, RR-image quality assessment techniques such as those in [16], [17] and [19] have used mask operations to extract structure information.

### 7.1.2   *Frequency-based methods*

Frequency domain-based features have been used in many multimedia applications. Discrete cosine transformation (DCT) coefficient and wavelet coefficient are the two most significant and widely used feature extraction domains in frequency transformation. Wavelet divides signal into multiple frequency components and analyzes these components with different scales. On the other hand, DCT [22] represents a finite signal into a sum of cosine functions oscillating at different frequencies. Researchers have utilized wavelet and DCT for RR-based quality estimation of an image and

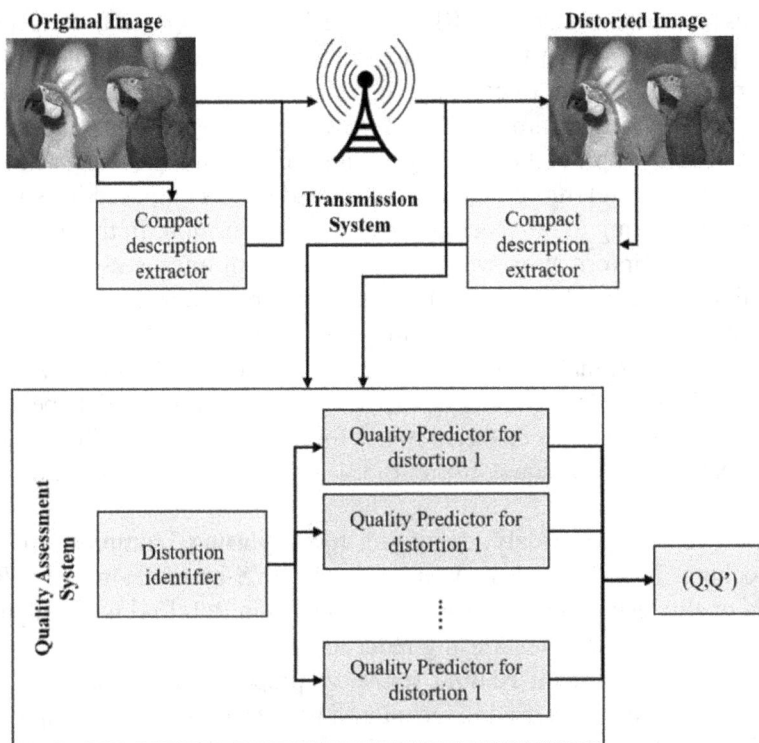

**Figure 7.3.** The two-layer overview of RR quality assessment system based on mask-based method [18].

a video. We have discussed techniques based on wavelet and DCT in the following.

Discrete wavelet transform (DWT) is used in wavelet-based watermarking techniques such as those in Refs. [23], [24] and [25]. At the sender's side, extracted features are inserted into an original image using the watermarking wavelet technique. At the receiver's end, embedded features from the transmitted image are extracted and compared with features extracted at the sender's side for quality estimation [26]. Natural image statistics (features for RR-based quality estimation of an image and a video) have been computed using wavelet domain with Kullback—Leibler deviations in technique proposed in Ref. [27]. In another RR-image quality assessment method [28], RR features of an image and video are extracted in wavelet-domain using multi-scale geometric analysis, contrast sensitivity function and Weber's law of just noticeable difference. Color stereoscopic images have been evaluated in RR-based techniques proposed in Refs. [29], [30], [31] and [32]. In the first step, these techniques used color image disparity measurement to compute disparity map of a reference (original) image and a distorted image. The disparity measurement is achieved by tensor structured properties and eigenvalues of stereoscopic images. In the second step, the reference image and distorted are decomposed into multiple channels using multi-spectral wavelet decomposition. In the final step, visual feature information was obtained from both images using contrast sensitivity function filtering. All features obtained from the above three steps are fused for the reference image and the distorted image for RR-based quality estimation of stereoscopic multimedia. The low-level features (such as an edge-pattern map) have been used to differentiate between a reference image and a distorted image for quality estimation. These low-level features in techniques [8] and [33] are computed using multi-scale wavelet transformation.

DCT has been widely employed for denoising, compression and deblocking. In addition, it is used to extract HVS-based features for RR-based quality estimation. A technique proposed in Ref. [34] has computed distributions of coefficients using reduced DCT sub-band for RR perceptual quality assessment. Natural statistical properties of images in DCT domain [35] have been used in techniques presented in Refs. [34] and [36] for RR-based assessment. In the technique in Ref. [37], the magnitude and phase of DCT have been computed for a reference image and a distorted image to assess RR-based quality estimation. In another technique [38], inter-sub-band and intra-sub-band statistical features have been computed in DCT domain for RR-based quality assessment.

### 7.1.3   *Three-dimensional (3D)-based method*

Nowadays, 3D object recognition has become an active research topic due to its advantages over 2D recognition and the availability of low-cost 3D cameras, e.g., Microsoft Kinect and Intel RealSense [40–42]. As the demand for 3D data increases, the quality of 3D is a major concern. A technique proposed in Ref. [43] has computed the normalized coefficient of luminance and map disparity of 3D images in the contourlet sub-band ( [44], [39] and [45]) using Gaussian scale mixture model. These computed features have been computed for reference and distorted images. Further, feature similarity index is used to estimate the similarity between features of the reference and the distorted images for quality assessment. Figure 7.4

(a)

(b)

**Figure 7.4.** The stereo images (a) symmetrically and (b) asymmetrically compressed by HEVC. [39]

shows the original stereo images (further used to construct 3D images) and their compressed images. The method in Ref. [39] estimates quality of these stereo images for providing quality 3D images.

Color and depth information have been utilized in RR-based techniques proposed in Refs. 46, 47 and [48] to estimate quality of a 3D video. A technique presented in Ref. [49] is able to measure degradation in a 3D video due to compression and transmission by extracting depth and color features from 3D video. It is challenging to facilitate good quality 3D video for many multimedia-based applications, online 3D video streaming quality measurement and quality of online services measurement. Therefore, quality assessment of 3D data would be the potential future direction.

# References

[1]  B. Ciubotaru, G.-M. Muntean, and G. Ghinea, "Objective assessment of region of interest-aware adaptive multimedia streaming quality," *IEEE Transactions on Broadcasting*, 55(2), 202–212, 2009.

[2]    S. Winkler, A. Sharma, and D. McNally, "Perceptual video quality and blockiness metrics for multimedia streaming applications," in *Proceedings of the International Symposium on Wireless Personal Multimedia Communications*, 2001, p. 547–552.

[3]    M. Shahid, A. Rossholm, B. Lövström, and H.-J. Zepernick, "No-reference image and video quality assessment: a classification and review of recent approaches," *EURASIP Journal on Image and Video Processing*, 2014, 1–32, 2014.

[4]    T. Zhu and L. Karam, "A no-reference objective image quality metric based on perceptually weighted local noise," *EURASIP Journal on Image and Video Processing*, 5, 01 2014.

[5]    W. Xue and X. Mou, "Reduced reference image quality assessment based on weibull statistics," in *2010 Second International Workshop on Quality of Multimedia Experience (QoMEX)*, 2010, pp. 1–6.

[6]    S. Winkler and P. Mohandas, "The evolution of video quality measurement: From psnr to hybrid metrics," *IEEE Transactions on Broadcasting*, 54(3), 660–668, 2008.

[7]    J. Wu, W. Lin, G. Shi, and A. Liu, "Reduced-reference image quality assessment with visual information fidelity," *IEEE Transactions on Multimedia*, 15(7), 1700–1705, 2013.

[8]    M. Zhang, W. Xue, and X. Mou, "Reduced reference image quality assessment based on statistics of edge," *Proceedings of SPIE - The International Society for Optical Engineering*, vol. 8299, 01 2011.

[9]    Y. Xu, D. Liu, Y. Quan, and P. Le Callet, "Fractal analysis for reduced reference image quality assessment," *IEEE transactions on image processing*, 24, 03 2015.

[10]   K. Okarma and P. Lech, "A statistical reduced-reference approach to digital image quality assessment," *Computer Vision and Graphics*, 5337, 43–54, 05 2009.

[11]   L. Ma, S. Li, and K. N. Ngan, "Reduced-reference video quality assessment of compressed video sequences," *IEEE Transactions on Circuits and Systems for Video Technology*, 22(10), 1441–1456, 2012.

[12]   M. Carnec, P. Le Callet, and D. Barba, "An image quality assessment method based on perception of structural information," in *Proceedings of International Conference on Image Processing*, vol. 3, 2003, pp. III–185.

[13]   K. Gu, G. Zhai, X. Yang, and W. Zhang, "A new reduced-reference image quality assessment using structural degradation model," in *Proceedings of IEEE International Symposium on Circuits and Systems (ISCAS)*, 2013, pp. 1095–1098.

[14]   Q. Li and Z. Wang, "General-purpose reduced-reference image quality assessment based on perceptually and statistically motivated image representation," in *Proceedings of 15th IEEE International Conference on Image Processing*, 2008, pp. 1192–1195.

[15]    P. Paudyal, F. Battisti, and M. Carli, "Reduced reference quality assessment of light field images," *IEEE Transactions on Broadcasting*, 65(1), 152–165, 2019.

[16]    E. Kalatehjari and F. Yaghmaee, "Using structural information for reduced reference image quality assessment," in *Proceedings of 4th International Conference on Computer and Knowledge Engineering (ICCKE)*, 2014, pp. 537–541.

[17]    X. Lv and Z. J. Wang, "Reduced-reference image quality assessment based on perceptual image hashing," in *Proceedings of 16th IEEE International Conference on Image Processing (ICIP)*, 2009, pp. 4361–4364.

[18]    J. A. Redi, P. Gastaldo, I. Heynderickx, and R. Zunino, "Color distribution information for the reduced-reference assessment of perceived image quality," *IEEE Transactions on Circuits and Systems for Video Technology*, 20(12), 1757–1769, 2010.

[19]    I. P. Gunawan and M. Ghanbari, "Reduced-reference video quality assessment using discriminative local harmonic strength with motion consideration," *IEEE Transactions on Circuits and Systems for Video Technology*, 18(1), 71–83, 2008.

[20]    Q. Li and Z. Wang, "Reduced-reference image quality assessment using divisive normalization-based image representation," *IEEE Journal of Selected Topics in Signal Processing*, 3(2), 202–211, 2009.

[21]    S. Dost, S. Anwer, F. Saud, and M. Shabbir, "Outliers classification for mining evolutionary community using support vector machine and logistic regression on azure ml," in *2017 International Conference on Communication, Computing and Digital Systems (C-CODE)*, 2017, pp. 216–221.

[22]    N. Ahmed, T. Natarajan, and K. Rao, "Discrete cosine transform," *IEEE Transactions on Computers*, C-23(1), 90–93, 1974.

[23]    A. N. Avanaki, S. Sodagari, and A. Diyanat, "Reduced reference image quality assessment metric using optimized parameterized wavelet watermarking," in *2008 9th International Conference on Signal Processing*, 2008, pp. 868–871.

[24]    W. Lu, X. Gao, D. Tao, and X. Li, "A wavelet-based image quality assessment method," *International Journal of Wavelets, Multiresolution and Information Processing*, 06(04), 541–551, 2008.

[25]    S. Altous, M. K. Samee, and J. Götze, "Reduced reference image quality assessment for jpeg distortion," in *Proceedings of ELMAR-2011*, 2011, pp. 97–100.

[26]    M. H. Kayvanrad, S. Sodagari, A. N. Avanaki, and H. Ahmadi-Noubari, "Reduced reference watermark-based image transmission quality metric," in *Proceedings of 3rd International Symposium on Communications, Control and Signal Processing*, 2008, pp. 526–531.

[27]　J. R. Hershey and P. A. Olsen, "Approximating the kullback leibler divergence between gaussian mixture models," in *Proceedings of IEEE International Conference on Acoustics, Speech and Signal Processing - ICASSP '07*, 4, IV–317–IV–320, 2007.

[28]　X. Gao, W. Lu, D. Tao, and X. Li, "Image quality assessment based on multiscale geometric analysis," *IEEE Transactions on Image Processing*, 18(7), 1409–1423, 2009.

[29]　W. Zhou, G. Jiang, M. Yu, F. Shao, and Z. Peng, "Reduced-reference stereoscopic image quality assessment based on view and disparity zero-watermarks," *Signal Processing: Image Communication*, vol. 29, 01 2013.

[30]　F. Qi, D. Zhao, and W. Gao, "Reduced reference stereoscopic image quality assessment based on binocular perceptual information," *IEEE Transactions on Multimedia*, 17(12), 2338–2344, 2015.

[31]　P. Campisi, P. Le Callet, and E. Marini, "Stereoscopic images quality assessment," in *Proceedings of 15th European Signal Processing Conference*, 2007, pp. 2110–2114.

[32]　A. Maalouf and M.-C. Larabi, "Cyclop: A stereo color image quality assessment metric," in *Proceedings of IEEE International Conference on Acoustics, Speech and Signal Processing (ICASSP)*, 2011, pp. 1161–1164.

[33]　J. Galbally, S. Marcel, and J. Fierrez, "Image quality assessment for fake biometric detection: Application to iris, fingerprint, and face recognition," *IEEE Transactions on Image Processing*, 23(2), 710–724, 2014.

[34]　M. Ul Haque, M. Qadri, and N. Siddiqui, "Reduced reference blockiness and blurriness meter for image quality assessment," *Imaging Science Journal The*, 63, 02 2015.

[35]　M. Saad, A. Bovik, and C. Charrier, "Blind image quality assessment: A natural scene statistics approach in the DCT domain," *IEEE Transactions on Image Processing*, 21(8), 3339–3352, 2012.

[36]　T. Kusuma and H.-J. Zepernick, "A reduced-reference perceptual quality metric for in-service image quality assessment," in *SympoTIC'03. Joint 1st Workshop on Mobile Future and Symposium on Trends in Communications*, 2003, pp. 71–74.

[37]　L. Ma, S. Li, and K. N. Ngan, "Reduced-reference image quality assessment in reorganized dct domain," *Signal Processing: Image Communication*, 28(8), 884–902, 2013.

[38]　M. Narwaria, W. Lin, I. V. McLoughlin, S. Emmanuel, and L.-T. Chia, "Fourier transform-based scalable image quality measure," *IEEE Transactions on Image Processing*, 21(8), 3364–3377, 2012.

[39]　Y.-H. Lin and J.-L. Wu, "Quality assessment of stereoscopic 3D image compression by binocular integration behaviors," *IEEE Transactions on Image Processing*, 23(4), 1527–1542, 2014.

[40] R. B. Rusu and S. Cousins, "3D is here: Point cloud library (PCL)," in *Proceedings of IEEE International Conference on Robotics and Automation*, 2011, pp. 1–4.

[41] A. Aldoma, F. Tombari, L. Di Stefano, and M. Vincze, "A global hypotheses verification method for 3D object recognition," in *Proceedings of European Conference on Computer Vision*, 2012, pp. 511–524.

[42] M. Kaiser, X. Xu, B. Kwolek, S. Sural, and G. Rigoll, "Towards using covariance matrix pyramids as salient point descriptors in 3D point clouds," *Neurocomputing*, 120, 101–112, 2013.

[43] X. Wang, Q. Liu, R. Wang, and Z. Chen, "Natural image statistics based 3d reduced reference image quality assessment in contourlet domain," *Neurocomputing*, 151, 683–691, 2015.

[44] W. T. Sheng, G. Xin-bo, L. Wen, L. Guang-dong, "A new method for reduced-reference image quality assessment," *Journal of Xidian University*, 35(1), 101–103, 2008.

[45] X. Wang, G. Jiang, and M. Yu, "Reduced reference image quality assessment based on contourlet domain and natural image statistics," 2009.

[46] C. T. Hewage and M. G. Martini, "Reduced-reference quality assessment for 3D video compression and transmission," *IEEE Transactions on Consumer Electronics*, 57(3), 1185–1193, 2011.

[47] A. Mittal, A. K. Moorthy, J. Ghosh, and A. C. Bovik, "Algorithmic assessment of 3d quality of experience for images and videos," in *Proceedings of Digital Signal Processing and Signal Processing Education Meeting (DSP/SPE)*, 2011, pp. 338–343.

[48] C. T. E. R. Hewage and M. G. Martini, "Reduced-reference quality metric for 3d depth map transmission," in *Proceedings of 3DTV-Conference: The True Vision - Capture, Transmission and Display of 3D Video*, 2010, pp. 1–4.

[49] A. Albonico, G. Valenzise, M. Naccari, M. Tagliasacchi, and S. Tubaro, "A reduced-reference video structural similarity metric based on no-reference estimation of channel-induced distortion," in *Proceedings of IEEE International Conference on Acoustics, Speech and Signal Processing*, 2009, pp. 1857–1860.

# Chapter 8

# No Reference Image Quality Assessment (NR-IQA)

NR-image quality assessment techniques do not need any reference image or features for estimation of quality of a distorted image. In literature, NR-image quality assessment techniques have been categorized into two categories: distortion-specific NR-image quality assessment and general-purpose NR-image quality assessment. A summary of categorization is shown in Figure 8.1 [1]. Distortion NR-image quality assessment techniques have used prior knowledge of distortion for quality estimation. Some distortion-specific techniques are as follows. Edge information and pixel distortion-based features have been used in Ref. [2] to estimate JP2K distortion. Blocking artifact has been assessed in the technique in Ref. [3] using the power of a blocky signal, in the technique in Ref. [4], with the use of discrete cosine transform (DCT), and in the technique in Ref. [5], using block-edge impairment metric. A technique proposed in Ref. [6] has applied machine learning on features including block boundary, image corners and color-changing properties for estimating blocking artifact. A technique proposed in Ref. [7] has used sharpness distribution of the gradient profile and visibility regions of gradient profiles to assess blocking and ringing artifacts, respectively. Discrete Tchebichef moments have been utilized in Ref. [8] to estimate blur in an image. Another technique [9] has computed edge spreading to assess bur. A combination of natural scenes statistics (NSS) model multi-resolution decomposition method has been utilized in Ref. [10] for blur estimation. Riesz transformation and bidimensional empirical mode decomposition methods have been used

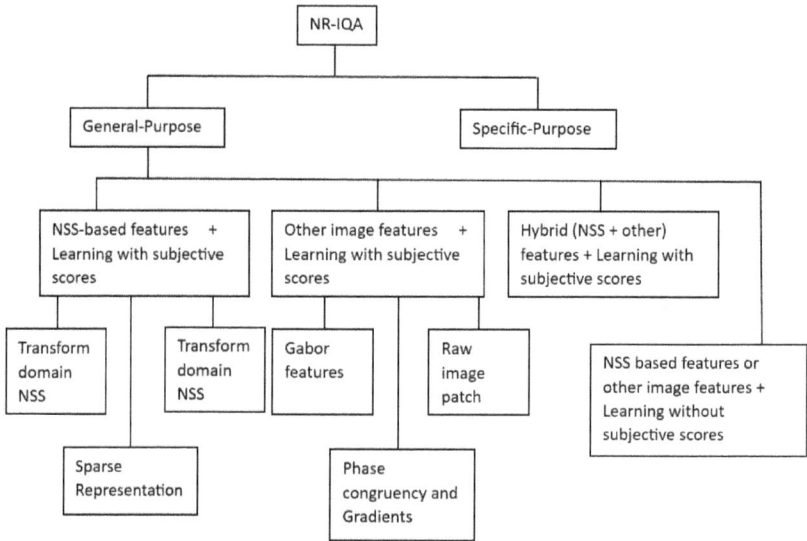

**Figure 8.1.** Different types of NR-image quality assessment techniques.

in Ref. [11] to estimate noise in an image. Multiple NSS features have been computed and used in support vector regression (SVR) to estimate contrast-based quality in Ref. [12]. Tchebichef moments (such as blur estimation) are also used by the DBIQ technique [13] to assess the quality of a deblocked image.

Multiple distortions are measured in the general purpose quality assessment category. A technique proposed in Ref. [14] has classified the distortion using distorted image statistics (DIS) features by applying support vector machine (SVM) and quality of classified distortion is measured using probability computed by SVM. In the BLIINDS technique [15], NSS features in the DCT domain, and the DIIVINE technique [16], NSS features in wavelet domain have been explored to estimate quality of an image. Another technique called BRISQUE [17] has used NSS features in spatial domain to estimate quality. A training-based technique which is presented in Ref. [18] has learned blur or noise statistics, complex wavelet domain statistics and distortion texture statistics for quality estimation. A technique proposed in Ref. [19] has divided an image into patches and then Gabor wavelet-based features have been extracted to form visual codebooks. Further, SVM is applied to estimate the quality. An interesting brain theory (based on HVS and free energy) is utilized in the training-based NR-image quality assessment technique [20]. General purpose techniques no doubt estimate more distortion than signal purpose NR-image quality assessment. However, they demand heavy training based on a human score of distorted images (subjective score). Therefore, their scope is limited to distortions provided for training [1].

In literature, a few techniques have been proposed that have not dependent on the subjective score for the training. A technique called quality-aware clustering (QAC) proposed in Ref. [21] has used a reference image and its four distorted images to train the NR-image quality assessment model. A technique proposed in Ref. [22] has used NSS features for training. Another NR-image quality assessment technique proposed in Ref. [23] has utilized entropy, mean phase congruency and mean gradient magnitude features for the training. The multivariate Gaussian model (MVG) has been explored in Ref. [24] to train NR-image quality assessment model to estimate quality. All the techniques mentioned above do not need a subjective score for the training. However, their training depends on reference and/or distorted images [1].

A few NR-image quality assessment techniques have been presented that do not require any training to estimate image quality. BIQES [1], SA-SS [25], SA-ZC [25] and NOMDA [26] are some techniques proposed in the literature that are independent of training. In BIQES [1], first, an image is represented using multiple scales and then information loss over these scales has been estimated for quality assessment. Structural activity (SA) information has been utilized in Ref. [25] and multiple differences of Gaussian images have been computed in Ref. [26] to estimate training-free quality. For applications where a reference image is not available, NR-image quality assessment techniques are preferred over FR-image quality assessment and RR-image quality assessment. Also, if an NR-image quality assessment is training-free, its applicability is increased in a real-time situation where subjective or original/distorted samples are unavailable for training. In the following section, we explore some training-free NR-image quality assessment techniques to gain an accurate and deep understanding.

## 8.1 Training free NR-image Quality Assessment Techniques

In this section, in detail, we discuss some NR-image quality assessment techniques for different distortions (such as contrast, illumination, noise and deblocked images). We present an experimental evaluation of NR-image quality assessment techniques on widely used databases which helps readers understand the research aspects of training-free NR-image quality assessment techniques.

### 8.1.1   *Contrast-based quality assessment*

Low contrast and high contrast are the cause of poor-quality images. A technique named contrast-based quality (CBQ) presented in Ref. [27] has estimated quality of an image degraded due to poor contrast (low and high contrast). This technique has made use of local band-limited contrast [28] for quality assessment and it is defined at pixel $(x,y)$ as

$$c(x,y) = \frac{\lambda(x,y)}{\beta(x,y)} \tag{8.1}$$

where $\lambda(x,y) = f(x,y) * b(x,y)$ and $\beta(x,y) = f(x,y) * l(x,y)$. $f(x,y)$ is the image under consideration, $*$ is the convolution operator, $l(x,y)$ is a low-pass filter and $b(x,y)$ is a band-pass filter. This technique has utilized a concept of modeling band-limited contrast using the difference of Gaussians (DoG) as given in the following:

$$O(x,y) = C(x,y) - S(x,y) \tag{8.2}$$

where $C(x,y) = f(x,y) * G_1(x,y)$ and $S(x,y) = f(x,y) * G_2(x,y)$. $C(x,y)$ and $S(x,y)$ are the two images filtered by two Gaussian functions $G_1$ and $G_2$, respectively.

Gaussian functions mentioned-above $(G_1(x,y), G_2(x,y))$ are differ by their standard deviation $(\sigma)$ values. In Equation 8.2, $O(x,y)$ is the image obtained by band-pass filter. Therefore, local band-limited contrast can be alternatively defined as

$$c(x,y) = \frac{O(x,y)}{S(x,y)} \tag{8.3}$$

This technique also has included multiple spatial frequencies [30] as an HVS concept to compute multiple band-limited contrasts for various frequencies to estimate image quality. To model multiple spatial frequencies, this technique has generated multiple standard deviation values at multiple frequencies for Gaussian function $G_1(x,y)$ as follows:

$$\sigma_1^i = \frac{2log(\frac{1}{L^2})}{f_i^2(1-L^2)}, f_i = \frac{3\pi}{80}, \frac{6\pi}{80}, \ldots, \frac{72\pi}{80} \tag{8.4}$$

where $\sigma_1^i$ is the standard deviation computed at $i-th$ frequency for the first Gaussian function and $f_i \in [0,\pi]$ represents the frequency for which the passband of the underlying band-pass filter peaks [31]. And $\sigma_2^i$ is computed as $\sigma_2^i = L * \sigma_1^i$, where $L$ ($> 1$) is a constant that gives $\sigma_2^i$ larger than $\sigma_1^i$. This technique has obtained a low-pass filtered image $S^i$ and a band-pass filtered image $O^i$ for each pair $\sigma_1^i$, $\sigma_2^i$. Further, this technique has computed the final contrast-based quality ($Q_C$) of an image using their proposed block-level binary thresholding method. $Q_C$ ranges between 0 and 1, where 0 depicts the lowest quality and 1 indicates the highest quality in the image. Figure 8.2 represents the results obtained by this technique. As the contrast of images in the figure increases, the quality score ($Q_C$) estimated by this technique also increases and vice versa. This shows the robustness of this technique. Another example is shown in Figure 8.3. Here, we can also see that as contrast decreases in natural images, the obtained quality score decreases.

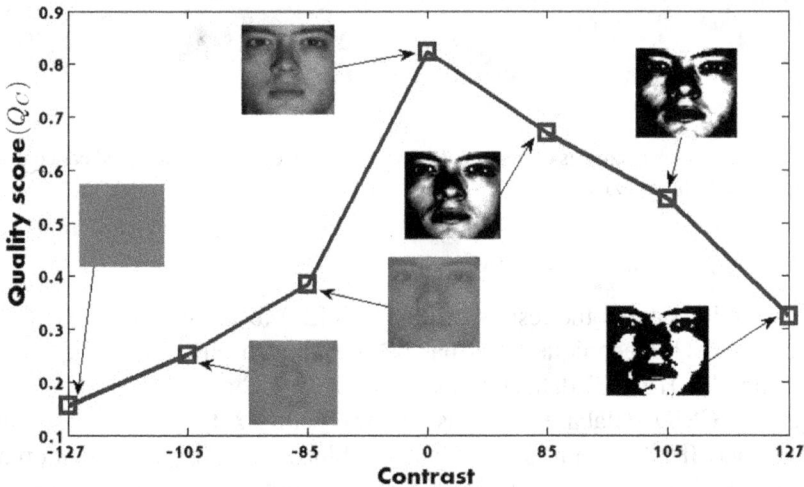

**Figure 8.2.** Low and high contrast images with their corresponding contrast-based estimated quality.

(a) $Q_C = 0.71$                    (b) $Q_C = 0.61$

(c) $Q_C = 0.26$                    (d) $Q_C = 0.13$

**Figure 8.3.**   Few examples of poor contrast images with estimated quality where contrast decreasing from (a) to (d) [29].

### 8.1.1.1    *Experimental results*

This section covers the results of the CBQ [27] technique on well-known databases. The technique CBQ has been evaluated on CSIQ [29], TID08 [32] and TID13 [33] databases using SROCC, CC and RMSE evaluation metrics. CSIQ database contains images with six types of distortions: JPEG and JP2K compressions, Gaussian blurring, additive Gaussian pink noise, additive Gaussian white noise and global contrast decrements. TID08 and TID13 databases contain images with distortions of 17 and 25 types, respectively. Distortions in these databases include different types of noises, blur, denoising, JPEG and JP2K compressions, the transmission of JPEG, JP2K compression with errors, contrast changes, luminance and some local distortions. To test the CBQ, we have considered all images from these databases distorted due to contrast. The description of all databases is summarized in Table 8.1.

The performance of the CBQ technique is compared with several other image quality assessment techniques, such as FSIM [34], PSNR [35], SSIM [35], NIQE [24], BRISQUE [17], BIQI [14], BIQES [1], NOMDA [26] and SISBLIM [36]. The techniques proposed in FSIM [34], PSNR [35] and SSIM [35] are full-reference techniques. The techniques presented in NIQE [24], BRISQUE [17] and BIQI [14] are training-based no-reference image quality assessment techniques, while techniques in BIQES [1], NOMDA [26] and SISBLIM [36] are no-training, no-reference-based image quality assessment techniques. Table 8.2 compares the CBQ technique with existing image quality assessment techniques for contrast distortion on CSIQ database. Table 8.3 shows the comparison

**Table 8.1.** Description of databases for contrast distortion.

| Database | Number of images |
|----------|------------------|
| CSIQ [29] | 116 |
| TID08 [32] | 100 |
| TID13 [33] | 125 |

**Table 8.2.** Result comparison with existing image quality assessment techniques for contrast distortion on CSIQ database.

| Quality measure | Type | Training | SROCC | CC | RMSE |
|-----------------|------|----------|-------|-----|------|
| FSIM [34] | FR | No | 0.9351 | 0.8662 | 0.0906 |
| PSNR [35] | FR | No | 0.8889 | 0.8988 | 0.0738 |
| SSIM [35] | FR | No | 0.822 | 0.8256 | 0.095 |
| NIQE [24] | NR | Yes | 0.2778 | 0.2572 | 0.1627 |
| BRISQUE [17] | NR | Yes | 0.0563 | 0.2244 | 0.1641 |
| BIQI [14] | NR | Yes | 0.5851 | 0.5509 | 0.1405 |
| NOMDA [26] | NR | No | 0.4168 | 0.3756 | 0.1531 |
| SISBLIM [36] | NR | No | 0.5403 | 0.5125 | 0.1446 |
| BIQES [1] | NR | No | 0.5403 | 0.5125 | 0.1446 |
| CBQ | NR | No | **0.8378** | **0.7928** | **0.1026** |

**Table 8.3.** Result comparison with existing image quality assessment techniques for contrast distortion on TID08 database.

| Quality measure | Type | Training | SROCC | CC | RMSE |
|---|---|---|---|---|---|
| FSIM [34] | FR | No | 0.6481 | 0.729 | 0.8375 |
| PSNR [35] | FR | No | 0.6126 | 0.6043 | 0.9749 |
| SSIM [35] | FR | No | 0.5232 | 0.4879 | 1.0681 |
| NIQE [24] | NR | Yes | 0.2084 | 0.1097 | 1.2162 |
| BRISQUE [17] | NR | Yes | 0.0382 | 0.023 | 1.2233 |
| BIQI [14] | NR | Yes | 0.5604 | 0.4492 | 1.0932 |
| NOMDA [26] | NR | No | 0.2614 | 0.2828 | 1.181 |
| SISBLIM [36] | NR | No | 0.4213 | 0.4346 | 1.102 |
| BIQES [1] | NR | No | 0.6271 | 0.4188 | 1.1111 |
| CBQ | NR | No | **0.7258** | **0.7660** | **0.7866** |

**Table 8.4.** Result comparison with existing image quality assessment techniques for contrast distortion on TID13 database.

| Quality measure | Type | Training | SROCC | CC | RMSE |
|---|---|---|---|---|---|
| FSIM [34] | FR | No | 0.4686 | 0.6869 | 0.8816 |
| PSNR [35] | FR | No | 0.4608 | 0.4816 | 1.0631 |
| SSIM [35] | FR | No | 0.3686 | 0.4647 | 1.0742 |
| NIQE [24] | NR | Yes | 0.1626 | 0.0671 | 1.2103 |
| BRISQUE [17] | NR | Yes | 0.0236 | 0.0134 | 1.213 |
| BIQI [14] | NR | Yes | 0.5045 | 0.3936 | 1.1152 |
| NOMDA [26] | NR | No | 0.2381 | 0.2294 | 1.1807 |
| SISBLIM [36] | NR | No | 0.399 | 0.3872 | 1.1185 |
| BIQES [1] | NR | No | 0.2628 | 0.1455 | 1.2002 |
| CBQ | NR | No | **0.7151** | **0.7271** | **0.8328** |

between the CBQ technique and existing image quality assessment techniques on TID08 database. Table 8.4 shows comparison results for the TID13 database. We can see that the CBQ technique has outperformed all state-of-the-art techniques for all databases.

## 8.1.2 *Illumination-based quality assessment*

A technique named illumination-based quality (IBQ) proposed in [37] has estimated quality based on illumination in an image. This technique has computed quality based on identifying bright and dark regions using a code at location $c = (x_c, y_c)$ (provided $I_c < T_L$ or $I_c > T_H$) as given in the following:

$$code(x_c, y_c) = \sum_{(x_c,y_c)\in win} \frac{E(I_p - I_c)}{(N-1)} \tag{8.5}$$

where $T_L$ (low threshold) and $T_H$ (high threshold) are used to locate dark and bright regions, respectively. A window *win* is located at $(x_c, y_c)$ with number of pixels $N$. $I_c$ is the intensity of center pixel for which code is to be computed and $I_c$ is the intensity of $p$-th neighbor of center pixel. A function $E(.)$ in Equation 8.5 is given as follows:

$$E(D) = \begin{cases} 1, & \text{if } abs(D) > \varepsilon \\ 0, & \text{otherwise} \end{cases} \tag{8.6}$$

The function $E(.)$ assigns value 1 when the difference between $I_p$ and $I_c$ is greater than a threshold $\varepsilon$. Otherwise, it assigns a value of 0. The final quality based on illumination is estimated using computed codes for an image (having $W$ width and $H$ height) as follows:

$$Q_I = \frac{\sum_{i=1}^{W} \sum_{j=1}^{H} code(x_i, y_i)}{W * H} \tag{8.7}$$

Figure 8.4 shows some natural images affected by poor illumination and their estimated quality. This technique has estimated low-quality scores for images as shown in Figures 8.4(a)–(c) (poor quality images) and estimated higher-quality scores for the image as shown in Figure 8.4(d) (comparatively better than other images).

### 8.1.2.1 *Experimental results*

The IBQ technique has been evaluated on 100 images of different sizes ($720 \times 480$ to $3264 \times 2448$) obtained from the Wang database [38] using evaluation metrics, such as SROCC, CC and RMSE. The performance of the IBQ technique is compared with several other image quality assessment techniques, such as NIQE [24], BRISQUE [17], BIQI [14], BIQES [1], NOMDA [26] and SISBLIM [36]. Table 8.5 presents a performance comparison of the IBQ technique with existing image quality assessment techniques for images of the Wang database.

It is observed from the table that the IBQ technique outperforms all training-based no-reference image quality assessment techniques, such as

(a) $Q_I = 0.2328$

(b) $Q_I = 0.4348$

(c) $Q_I = 0.6287$

(d) $Q_I = 0.7312$

**Figure 8.4.**   Example of a few images with poor illumination with their estimated quality scores [37].

NIQE [24], BRISQUE [17] and BIQI [14], as well as no-training, no-reference-based image quality assessment techniques, such as NOMDA [26], SISBLIM [36] and BIQES [1], for all three evaluation metrics (i.e., SROCC, CC and RMSE).

### 8.1.3   *Noise-based quality assessment*

A technique named noise-based quality (NBQ) proposed in Ref. [39] has used two HVS concepts: multiple spatial frequency channels and centre-surround receptive field to estimate noise-based image quality. This technique has modeled centre-surround receptive field using DoG images computed at different frequencies [27]. Further, this technique extracts

**Table 8.5.** Performance comparison of illumination-based quality estimation technique with other existing image quality assessment techniques on Wang database.

| Quality measure | Type | Training | SROCC | CC | RMSE |
|---|---|---|---|---|---|
| NIQE [24] | NR | Yes | 0.2988 | 0.1691 | 16.7486 |
| BRISQUE [17] | NR | Yes | 0.3064 | 0.2603 | 16.4077 |
| BIQI [14] | NR | Yes | 0.0647 | 0.1084 | 16.8931 |
| NOMDA [26] | NR | No | 0.4923 | 0.4532 | 15.1478 |
| SISBLIM [36] | NR | No | 0.0693 | 0.0367 | 16.9819 |
| BIQES [1] | NR | No | 0.2469 | 0.1577 | 16.7806 |
| IBQ | NR | No | **0.7099** | **0.7273** | **11.6625** |

features from generated DoG images using singular value decomposition (SVD) [40] to estimate the image's quality. This technique has utilized lower singular values obtained by SVD as they are sensitive to noise variation in an image. This technique has computed quality ($Q_N$) as follows:

$$Q_N = \frac{1}{n}\sum_{i=1}^{n} T^i(v) \tag{8.8}$$

where $v$ indicates a vector that contains mean of lower singular values of different blocks of a DoG image (computed for $i$-th frequency) and $n$ represents the number of frequencies. This technique has again computed mean of lower values of vector $\mathbf{v}$ using function $T$. Figure 8.5 shows a few noisy images (with standard deviation) with their estimated quality score ($Q_N$). It can be seen that this technique can estimate the presence of noise in the image. For example, the lowest standard deviation (lowest level of noise) has the lowest quality score and vice versa.

### 8.1.3.1 Experimental results

The NBQ technique has been evaluated on LIVE [41], CSIQ [29], TID08 [32] and SD-IVL [42] databases. These databases contain images of natural scenes, buildings, people, boats, animals, etc. To test the NBQ technique, we have considered all the images from the databases distorted due to white noise. All databases used in the evaluation are summarized in Table 8.6.

The performance of the NBQ technique is compared with several other NR-image quality assessment techniques, such as BIQI [14], BRISQUE

(a) Noise $\sigma = 0.0937$,
$Q_N = 11.22$

(b) Noise $\sigma = 0.1367$,
$Q_N = 16.36$

(c) Noise $\sigma = 1.0$,
$Q_N = 54.70$

(d) Noise $\sigma = 0.3125$,
$Q_N = 25.52$

(e) Noise $\sigma = 0.4218$,
$Q_N = 32.34$

(f) Noise $\sigma = 1.5$,
$Q_N = 55.59$

**Figure 8.5.** Quality score $Q_N$ for few noisy images [39].

**Table 8.6.** Description of databases for white noise distortion.

| Database | Number of images |
|---|---|
| LIVE [41] | 174 |
| CSIQ [29] | 150 |
| TID08 [32] | 100 |
| SD-IVL [42] | 200 |

[17], NIQE [24], SISBLIM [36], BIQES [1] and NOMDA [26]. The technique is also compared with FR-image quality assessment techniques, such as PSNR [35], FSIM [34] and SSIM [35]. Table 8.7 compares the NBQ technique with existing training and no-training-based image quality assessment techniques for white noise distortion on LIVE database.

**Table 8.7.** Result comparison with existing image quality assessment techniques for noise distortion on LIVE database.

| Quality measure | Type | Training | SROCC | CC | RMSE |
|---|---|---|---|---|---|
| PSNR [35] | FR | No | 0.9810 | 0.9587 | 8.1168 |
| FSIM [34] | FR | No | 0.9546 | 0.9523 | 8.4596 |
| SSIM [35] | FR | No | 0.9418 | 0.9512 | 8.6261 |
| NIQE [24] | NR | Yes | 0.9662 | 0.8441 | 11.3852 |
| BRISQUE [17] | NR | Yes | 0.9786 | 0.8884 | 10.0914 |
| BIQI [14] | NR | Yes | 0.9510 | 0.8773 | 10.3496 |
| NOMDA [26] | NR | No | 0.9786 | 0.9546 | 8.2463 |
| SISBLIM [36] | NR | No | 0.9541 | 0.9123 | 10.4578 |
| BIQES [1] | NR | No | 0.9694 | 0.9315 | 9.7505 |
| NBQ | NR | No | **0.9846** | **0.9205** | **9.9107** |

**Table 8.8.** Result comparison with existing image quality assessment techniques for noise distortion on CSIQ database.

| Quality measure | Type | Training | SROCC | CC | RMSE |
|---|---|---|---|---|---|
| PSNR [35] | FR | No | 0.9347 | 0.9428 | 0.0559 |
| FSIM [34] | FR | No | 0.9271 | 0.7713 | 0.1068 |
| SSIM [35] | FR | No | 0.8799 | 0.8622 | 0.085 |
| NIQE [24] | NR | Yes | 0.6668 | 0.6219 | 0.1314 |
| BRISQUE [17] | NR | Yes | 0.7743 | 0.7816 | 0.1047 |
| BIQI [14] | NR | Yes | 0.7938 | 0.8018 | 0.1003 |
| NOMDA [26] | NR | No | 0.8694 | 0.8427 | 0.0903 |
| SISBLIM [36] | NR | No | 0.7970 | 0.8099 | 0.0914 |
| BIQES [1] | NR | No | 0.6925 | 0.6971 | 0.1203 |
| NBQ | NR | No | **0.9213** | **0.9011** | **0.0728** |

In this case, it is seen that the NBQ technique is superior to all training and no-training-based image quality assessment techniques in terms of SROCC value. Table 8.8 shows comparison results on CSIQ database. The NBQ technique outperforms all existing techniques for all evaluation metrics in this case, except PSNR [35]. PSNR [35] gives slightly better performance than the NBQ technique; however, it must be noted that the PSNR [35] is a full-reference-based technique while the NBQ technique is no-reference based. Table 8.9 shows comparison results on TID08 database. The NBQ technique is better than all the existing techniques such as FSIM [34], SSIM [35], NIQE [24], BRISQUE [17], BIQI [14],

**Table 8.9.**   Result comparison with existing image quality assessment techniques for noise distortion on TID08 database.

| Quality measure | Type | Training | SROCC | CC | RMSE |
|---|---|---|---|---|---|
| PSNR [35] | FR | No | 0.9115 | 0.9327 | 0.2203 |
| FSIM [34] | FR | No | 0.8566 | 0.7828 | 0.3802 |
| SSIM [35] | FR | No | 0.7791 | 0.7531 | 0.5187 |
| NIQE [24] | NR | Yes | 0.5701 | 0.5282 | 1.2162 |
| BRISQUE [17] | NR | Yes | 0.7065 | 0.6767 | 0.4498 |
| BIQI [14] | NR | Yes | 0.6087 | 0.5765 | 0.4992 |
| NOMDA [26] | NR | No | 0.8459 | 0.8184 | 0.351 |
| SISBLIM [36] | NR | No | 0.7276 | 0.5668 | 0.5033 |
| BIQES [1] | NR | No | 0.7191 | 0.6806 | 0.4476 |
| NBQ | NR | No | **0.8551** | **0.8243** | **0.3459** |

**Table 8.10.**   Result comparison with existing image quality assessment techniques for noise distortion on SD-IVL database.

| Quality measure | Type | Training | SROCC | CC | RMSE |
|---|---|---|---|---|---|
| PSNR [35] | FR | No | 0.9734 | 0.9553 | 6.988 |
| FSIM [34] | FR | No | 0.9057 | 0.697 | 16.9447 |
| SSIM [35] | FR | No | 0.8259 | 0.7106 | 16.6245 |
| NIQE [24] | NR | Yes | 0.158 | 0.1796 | 22.4872 |
| BRISQUE [17] | NR | Yes | 0.3728 | 0.372 | 21.9332 |
| BIQI [14] | NR | Yes | 0.4752 | 0.4389 | 21.2319 |
| NOMDA [26] | NR | No | 0.8753 | 0.8383 | 12.8837 |
| SISBLIM [36] | NR | No | 0.4697 | 0.3647 | 22.0017 |
| BIQES [1] | NR | No | 0.418 | 0.4123 | 21.527 |
| NBQ | NR | No | **0.9194** | **0.8804** | **11.2063** |

NOMDA [26], SISBLIM [36] and BIQES [1] on TID08 database. Again, PSNR [35] performs slightly better than the NBQ technique in this case, too; however, as stated above, PSNR [35] is a full-reference-based technique, while the NBQ technique is no-reference based. Table 8.10 presents comparison results on SD-IVL database. In this case, also, the NBQ technique clearly outperforms all other image quality assessment techniques in terms of all evaluation metrics except PSNR [35]. The results of the PSNR [35] are slightly better than the NBQ technique in this case, too; however, as stated above, PSNR [35] needs a reference image to estimate the quality of a distorted image, while the NBQ technique is a no-reference-based technique. From Tables 8.7, 8.8, 8.9 and 8.10, we can see that the NBQ technique has achieved superior results and has outperformed the existing full-reference (except PSNR [35]) and no-reference (training and no-training)-based image quality assessment techniques. It is observed that PSNR [35] has performed the best; however, again, it is an FR-image quality assessment technique which needs a reference image, whereas the NBQ technique is based on no reference.

### 8.1.4 *Quality assessment of deblocked images*

A deblocked image mainly contains two distortions, *viz.* blocking artifacts and blur. The technique named quality assessment of deblocked images (QADI) [43] uses both of these distortions to evaluate a deblocked image's quality. It is observed that in a deblocked image, the quality of a smooth region is mainly affected due to blocking artifacts, whereas, in a textured region, it is mainly affected due to blur. The QADI technique estimates blocking artifacts and blur in the smooth and textured regions, respectively, to estimate the quality of a deblocked image. In the QADI technique, 2-D CWT is used to classify blocks of an image into smooth and textured regions. For this, the input image whose quality estimation is to be carried out is divided into blocks of size $N \times N$. After computation of 2-D CWT, an average value of 2-D CWT coefficients for each block is computed. Based on the obtained value, each block is classified into the smooth or the textured block using threshold $t$. If the average of 2-D CWT coefficients for a block is less than or equal to the threshold $t$, it represents a smooth block; otherwise, the block is classified as a textured block.

The QADI technique estimates blocking artifacts in the smooth blocks, as blockiness is the dominant distortion in smooth regions. To do this, a Canny edge detector is utilized and obtained edges, where more edges in the smooth block indicate the presence of blocking artifacts. The QADI technique estimates blur in the textured blocks as it is the dominant distortion found in the textured regions of a deblocked image. The extent to which blur is present in a textured block is determined by computing the correlation between it and its artificially generated blurred version. The weighted combination of blocking artifact score for the smooth regions and blur score of the textured regions gives the overall quality score.

#### 8.1.4.1 *Experimental results*

The performance evaluation of the QADI technique has been conducted on DBID database [13] using various evaluation metrics, such as SROCC,

**Table 8.11.** Comparison of the results of the QADI technique with existing NR-image quality assessment and FR-image quality assessment techniques for quality estimation of deblocked images on DBID database.

| Metric | Type | Training | SROCC | RMSE | CC |
|---|---|---|---|---|---|
| PSNR [35] | FR | No | 0.4299 | 0.9611 | 0.4018 |
| FSIM [34] | FR | No | 0.3213 | 1.0621 | 0.3044 |
| SSIM [35] | FR | No | 0.3867 | 1.0315 | 0.3214 |
| PSNR-B [44] | FR | No | 0.5279 | 0.9471 | 0.5068 |
| BIQI [14] | NR | Yes | 0.3170 | 1.0276 | 0.3537 |
| BRISQUE [17] | NR | Yes | 0.3089 | 1.0340 | 0.3381 |
| NIQE [24] | NR | Yes | 0.3818 | 0.9614 | 0.4840 |
| NOMDA [26] | NR | No | 0.3251 | 1.0341 | 0.3467 |
| SISBLIM [36] | NR | No | 0.2415 | 1.0378 | 0.3289 |
| BIQES [1] | NR | No | 0.4442 | 0.9828 | 0.4899 |
| DBIQ [13] | NR | No | 0.8600 | 0.5232 | 0.8793 |
| **QADI** | **NR** | No | **0.8617** | **0.5179** | **0.8613** |

RMSE and CC. The QADI technique has been compared with several NR-image quality assessment techniques, such as BIQI [14], BRISQUE [17], NIQE [24], SISBLIM [36], BIQES [1], NOMDA [26] and DBIQ [13] (specifically for deblocked images). The technique is also compared with FR-image quality assessment techniques, such as PSNR [35], FSIM [34], SSIM [35] and PSNR-B [44] (designed to estimate the quality of deblocked images). Comparison results are presented in Table 8.11. It can be seen from Table 8.11 that the QADI technique outperforms the existing state-of-the-art NR-image quality assessment and FR-image quality assessment techniques. It gives better results in terms of all the evaluation metrics (SROCC, RMSE and CC) than most existing techniques.

# References

[1] A. Saha and Q. Wu, "Utilizing image scales towards totally training free blind image quality assessment," *IEEE Transactions on Image Processing*, 24(6), 1879–1892, June 2015.

[2] Z. P. Sazzad, Y. Kawayoke, and Y. Horita, "No-reference image quality assessment for JPEG2000 based on spatial features," *Signal Processing: Image Communication*, 23(4), 257–268, 2008.

[3]   Z. Wang, A. Bovik, and B. Evan, "Blind measurement of blocking artifacts in images," in *Proceedings of International Conference on Image Processing (ICIP 2000)*, 3, pp. 981–984, 2000.

[4]   A. Bovik and S. Liu, "DCT-domain blind measurement of blocking artifacts in DCT-coded images," in *Proceedings of IEEE International Conference on Acoustics, Speech, and Signal Processing (ICASSP 2001)*, 3, 1725–1728, 2001.

[5]   H. Wu and M. Yuen, "A generalized block-edge impairment metric for video coding," *IEEE Signal Processing Letters*, 4(11), 317–320, 1997.

[6]   L. Li, W. Lin, and H. Zhu, "Learning structural regularity for evaluating blocking artifacts in jpeg images," *IEEE Signal Processing Letters*, 21(8), 918–922, 2014.

[7]   L. Liang, S. Wang, J. Chen, S. Ma, D. Zhao, and W. Gao, "No-reference perceptual image quality metric using gradient profiles for JPEG2000," *Signal Processing: Image Communication*, 25(7), 502–516, 2010.

[8]   L. Li, W. Lin, X. Wang, G. Yang, K. Bahrami, and A. C. Kot, "No-reference image blur assessment based on discrete orthogonal moments," *IEEE Transactions on Cybernetics*, 46(1), 39–50, 2016.

[9]   E. Ong, W. Lin, Z. Lu, X. Yang, S. Yao, F. Pan, L. Jiang, and F. Moschetti, "A no-reference quality metric for measuring image blur," in *Proceedings of Seventh International Symposium on Signal Processing and Its Applications*, 1, 469–472, 2003.

[10]  M.-J. Chen and A. C. Bovik, "No-reference image blur assessment using multiscale gradient," *EURASIP Journal on Image and Video Processing*, 2011(1), 1–11, 2011.

[11]  G. Yang, Y. Liao, Q. Zhang, D. Li, and W. Yang, "No-reference quality assessment of noise-distorted images based on frequency mapping," *IEEE Access*, PP(99), 1–1, 2017.

[12]  Y. Fang, K. Ma, Z. Wang, W. Lin, Z. Fang, and G. Zhai, "No-reference quality assessment of contrast-distorted images based on natural scene statistics," *IEEE Signal Processing Letters*, 22(7), 838–842, 2015.

[13]  L. Li, Y. Zhou, W. Lin, J. Wu, X. Zhang, and B. Chen, "No-reference quality assessment of deblocked images," *Neurocomputing*, 177, 572–584, 2016.

[14]  A. Moorthy and A. Bovik, "A two-step framework for constructing blind image quality indices," *IEEE Signal Processing Letters*, 17(5), 513–516, May 2010.

[15]  M. Saad, A. Bovik, and C. Charrier, "Blind image quality assessment: a natural scene statistics approach in the DCT domain," *IEEE Transactions on Image Processing*, 21(8), 3339–3352, 2012.

[16]  A. Moorthy and A. Bovik, "Blind image quality assessment: From natural scene statistics to perceptual quality," *IEEE Transactions on Image Processing,* 20(12), 3350–3364, 2011.

[17]  A. Mittal, A. Moorthy, and A. Bovik, "No-reference image quality assessment in the spatial domain," *IEEE Transactions on Image Processing,*, 21(12), 4695–4708, Dec 2012.

[18]  H. Tang, N. Joshi, and A. Kapoor, "Learning a blind measure of perceptual image quality," in *Proceedings of IEEE Conference on Computer Vision and Pattern Recognition (CVPR 2011).* IEEE, 2011, pp. 305–312.

[19]  P. Ye and D. Doermann, "No-reference image quality assessment using visual codebooks," *IEEE Transactions on Image Processing,* 21(7), 3129–3138, 2012.

[20]  K. Gu, G. Zhai, X. Yang, and W. Zhang, "Using free energy principle for blind image quality assessment," *IEEE Transactions on Multimedia,* 17(1), 50–63, 2015.

[21]  W. Xue, L. Zhang, and X. Mou, "Learning without human scores for blind image quality assessment," in *Proceedings of IEEE Conference on Computer Vision and Pattern Recognition (CVPR 2013),* June 2013, pp. 995–1002.

[22]  A. Mittal, G. Muralidhar, J. Ghosh, and A. Bovik, "Blind image quality assessment without human training using latent quality factors," *IEEE Signal Processing Letters,* 19(2), 75–78, 2012.

[23]  C. Li, Y. Ju, A. C. Bovik, X. Wu, and Q. Sang, "No-training, no-reference image quality index using perceptual features," *Optical Engineering,* 52(5), 057003–057003, 2013.

[24]  A. Mittal, R. Soundararajan, and A. Bovik, "Making a "completely blind" image quality analyzer," *Signal Processing Letters, IEEE,* 20(3), 209–212, 2013.

[25]  J. Zhang, T. M. Le, S. Ong, and T. Q. Nguyen, "No-reference image quality assessment using structural activity," *Signal Processing,* 91(11), 2575–2588, 2011.

[26]  P. Joshi and S. Prakash, "Retina inspired no-reference image quality assessment for blur and noise," *Multimedia Tools and Applications,* 76(18), 18871–18890, 2017.

[27]  P. Joshi and S. Prakash, "An efficient technique for image contrast enhancement using artificial bee colony," in *IEEE International Conference on Identity, Security and Behavior Analysis (ISBA 2015),* 2015, pp. 1–6.

[28]  E. Peli, "Contrast in complex images," *The Journal of the Optical Society of America A,* 7(10), 2032–2040, 1990.

[29]  E. C. Larson and D. M. Chandler, "Most apparent distortion: full-reference image quality assessment and the role of strategy," *Journal of Electronic Imaging,* 19(1), 011006(1–21), 2010.

[30]  R. L. DeValois and K. K. DeValois, (Eds.), *Spatial Vision*. Oxford University Press, Oxford, 1990.

[31]  D. Sen and S. K. Pal, "Automatic exact histogram specification for contrast enhancement and visual system based quantitative evaluation," *IEEE Transactions on Image Processing*, 20(5), 1211–1220, 2011.

[32]  A. Z. N. Ponomarenko, V. Lukin, K. Eziazarian, M. Carli, and F. Battisti, "TID2008— a database for evaluation of full reference image quality metrics," *Advance Modern Radioelectronics*, 10(4), 30–45, 2009.

[33]  N. Ponomarenko, L. Jin, O. Ieremeiev, V. Lukin, K. Egiazarian, J. Astola, B. Vozel, and Kacem, "Image database TID2013: Peculiarities, results and perspectives," *Signal Processing: Image Communication*, 30(C), 57–77, 2015.

[34]  L. Zhang, L. Zhang, X. Mou, and D. Zhang, "FSIM: A feature similarity index for image quality assessment," *IEEE Transactions on Image Processing*, 20(8), 2378–2386, 2011.

[35]  Z. Wang, A. Bovik, H. Sheikh, and E. Simoncelli, "Image quality assessment: from error visibility to structural similarity," *IEEE Transactions on Image Processing,*, 13(4), 600–612, 2004.

[36]  K. Gu, G. Zhai, X. Yang, and W. Zhang, "Hybrid no-reference quality metric for singly and multiply distorted images," *IEEE Transactions on Broadcasting*, 60(3), 555–567, 2014.

[37]  P. Joshi and S. Prakash, "Image enhancement with naturalness preservation," *Visual Computer*, 36(1), 71–83, 2020.

[38]  S. Wang, J. Zheng, H. M. Hu, and B. Li, "Naturalness preserved enhancement algorithm for non-uniform illumination images," *IEEE Transactions on Image Processing*, 22(9), 3538–3548, 2013.

[39]  P. Joshi and S. Prakash, "Nr-IQA for noise-affected images using singular value decomposition," *IET Signal Processing*, 13(2), 183–191, 2019.

[40]  G. H. Golub and C. F. V. Loan, *Matrix Computations*. JHU Press, Baltimore, MD, 1996.

[41]  L. C. H.R. Sheikh, Z.Wang and A. Bovik, "LIVE image quality assessment database release 2, LIVE, University of Texas at Austin, 2006. Available: http://live.ece.utexas.edu/research/quality."

[42]  S. Corchs, F. Gasparini, and R. Schettini, "No reference image quality classification for jpeg-distorted images," *Digital Signal Processing*, 30, 86–100, 2014.

[43]  P. Joshi, S. Prakash, and S. Rawat, "Continuous wavelet transform-based no-reference quality assessment of deblocked images," *The Visual Computer*, 34, 1739–1748, 2017.

[44]  C. Yim and A. C. Bovik, "Quality assessment of deblocked images," *IEEE Transactions on Image Processing*, 20(1), 88–98, 2011.

# Chapter 9

# Quality Aware Image Enhancement

Digital images play a significant role in human life due to their tremendous use in communications, monitoring, medical imaging, security and entertainment. However, the quality of images may degrade while passing through several operational stages, such as image acquisition, processing, transmission, compression and reconstruction [1]. For example, an image may be degraded by losing some data during transmission due to the limited bandwidth of the channel. Similarly, lossy compression may distort images by inserting blur and ringing artifacts. These distorted images can affect the performance of an image-based application. In the literature, several enhancement techniques are available to enhance poor-quality images. The existing techniques often blindly apply enhancement on images of different kinds without considering the nature and level of distortions present. For example, a quality unaware enhancement technique like the histogram equalization method blindly enhances an input good quality image where there is no need for the enhancement. It will lead to degradation of the input image by over-enhancement [2]. This degradation can be eliminated by estimating and utilizing the quality of an image before applying the enhancement. A few techniques have been developed in the literature to enhance the estimated quality of an image. We have categorized quality-aware image enhancement techniques into contrast-aware, illumination-aware and noise-aware quality enhancement.

# 9.1    Quality-Aware Contrast Enhancement

A technique proposed in Ref. [3] has enhanced contrast using artificial bee colony (ABC) algorithm [4]. First, this technique evaluates the contrast-based quality of an image and then the estimated quality is used as a fitness value of an image in the ABC algorithm to enhance poor-quality images. An efficient technique is proposed for automatic contrast enhancement in Ref. [5]. In order to estimate quality, phase congruency and statistics information of an image's histogram have been utilized to compute reduced-reference image quality metrics for contrast change (RIQMC). Further, RIQMC-based optimal histogram mapping (ROHIM) is used for automatic enhancement. An enhancement technique for finger vein images has been proposed in Ref. [6]. Using the estimated quality, this technique first classifies finger vein images into two classes high and low quality. Further, high and low-quality images have been enhanced separately using a single-scale Retinex filter with chromaticity preservation (SSRCP). To gain insight into quality-aware contrast enhancement, we discuss the ABC algorithm for quality-aware enhancement in the following subsection.

### 9.1.1    *Contrast enhancement using artificial bee colony (CEABC)*

This section presents a technique named CEABC, which is based on ABC algorithm for contrast enhancement [7]. This technique has proposed a direction constraint that helps artificial bees to search in the right direction. In addition, this technique has used contrast-based quality estimation as an objective function in ABC algorithm. Direction constraint states the presence of low or high contrast in an image. In ABC, first, all solutions are randomly initialized based on direction constraints. After initialization, the following steps are applied and repeated till stopping criteria are met for the enhancement [4]:

1. **Employed bee stage:** All solutions are updated in this step by the following equation:

$$v_{i,j} = x_{i,j} + \phi * (x_{i,j} - x_{k,j}) \qquad (9.1)$$

where $x_{i,j}$ is the current solution, $v_{i,j}$ is the updated solution, $\phi$ is a random number between -1 and 1 and $k$ is the random chosen solution from the population. For each newly generated solution, this technique

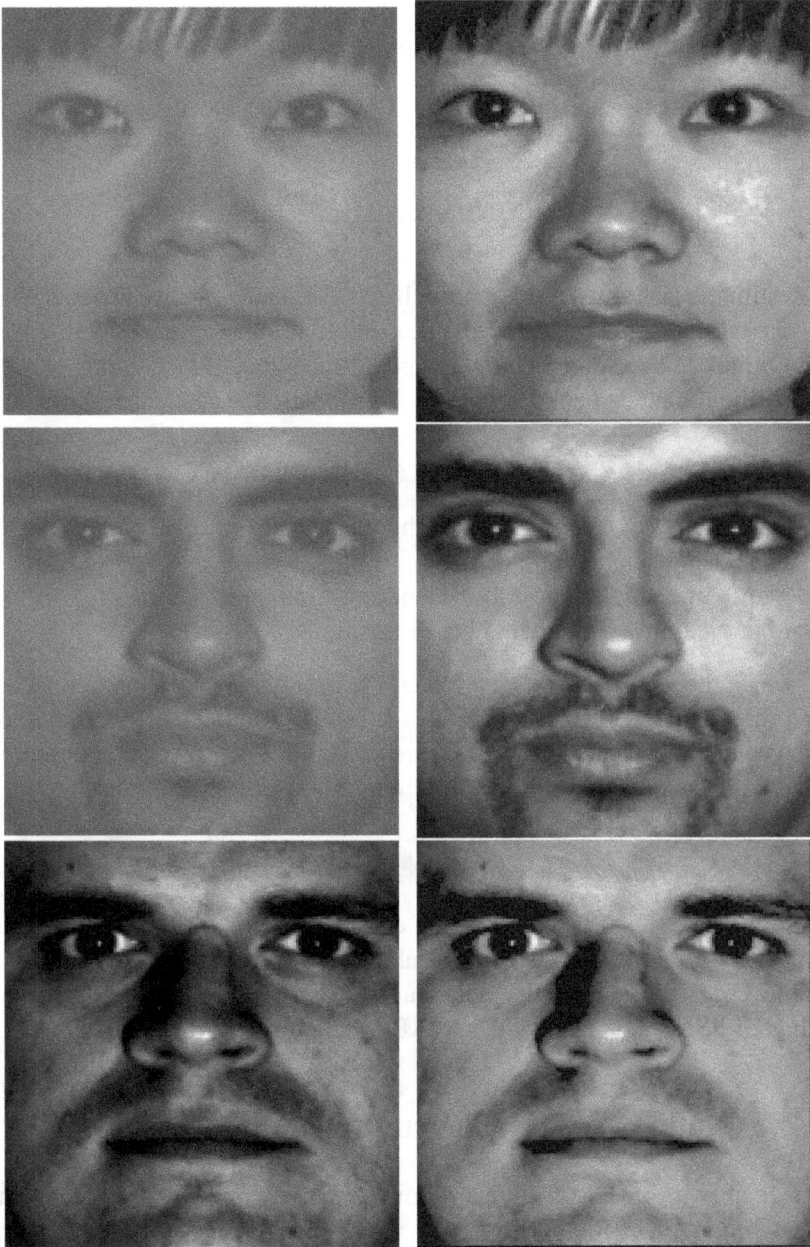

**Figure 9.1.** Column 1 shows poor contrast images and column 2 shows corresponding quality-aware enhanced images [7].

**Table 9.1.** Comparison of PSNR values for GA [9] and ABC [10] and CEABC technique.

| Database | GA [9] | ABC [10] | CEABC |
|----------|--------|----------|-------|
| Kodak database | 12.45 | 13.49 | 13.79 |
| Yale database | 12.05 | 12.14 | 13.16 |

estimates contrast-based quality. If this estimated quality is better than the current solution's, keep the new solution; otherwise, reject it.

2. **Onlooker bee stage:** This step updates neighbors of a solution chosen by the roulette wheel process [8]. The updation is given as follows:

$$v_{i,j} = x_{i,j} + \phi * (x_{i,j} - x_{c,j}) \qquad (9.2)$$

where $x_{c,j}$ is the solution chosen by the roulette wheel. The updation is done using neighbor $x_{i,j}$ of $x_{c,j}$.

3. **Scout bee stage:** These bees randomly initialize a solution it is not improved for a defined number of iterations.

Once the stopping criteria are met, the best quality solution is selected and the enhanced image is considered. A few examples of poor contrast images and their enhanced images are shown in Figure 9.1. It is observed that the visual quality improves after quality-aware enhancement.

### 9.1.2  *Experimental results*

Results of the CEABC technique are compared with GA [9] and ABC [10] on Kodak database [11] and Yale database [12] and the results are shown in Table 9.1. It is seen from the table that the CEABC technique has achieved a higher PSNR value than GA and ABC techniques for both databases.

## 9.2  Quality-Aware Illumination Enhancement

The quality-aware adaptive logarithmic enhancement (ALE) technique has been proposed in Ref. [13] to add a suitable amount of brightness for enhancing an image. An automatic fingerprint image enhancement technique has been proposed in Ref. [14]. This technique has classified fingerprint images into five categories for automatic parameter selection in contrast-limited adaptive histogram equalization (CLAHE) to enhance fingerprint images. In further subsection, we understand illumination-based quality enhancement technique [13] for a better understanding of quality-aware illumination enhancement. This technique has used an adaptive logarithm (for dynamic range compression) for enhancement computed based on the illumination-based quality of an image.

## 9.2.1 *Illumination enhancement*

The quality-aware enhancement technique named ALE proposed in Ref. [13] has used an adaptive logarithm (for dynamic range compression) computed based on the illumination-based quality to enhance non-uniform illumination in an image. An adaptive logarithm is used to add a sufficient amount of brightness in dark images for naturalness preservation, and it is given as

$$I_{ALE} = log(I + e) \qquad (9.3)$$

where $I$ is an input image, and the amount of brightness insertion to dark images depends on value $e$. Figure 9.2 shows the importance of choosing a proper value for $e$. Figure 9.2 (a) is the input image. If we blindly choose $e$, the proper enhancement for a different level of poor-quality images cannot be achieved. Therefore, this technique first estimates illumination-based quality and then sets a value for $e$ to get proper enhancement as follows:

$$e = \begin{cases} e_1, & \text{if } Q_d \geq \tau_1 \\ e_2, & \text{if } Q_d \geq \tau_2 \text{ and } Q_d < \tau_1 \\ e_3, & \text{if } Q_d \geq \tau_3 \text{ and } Q_d < \tau_2 \\ e_4, & \text{if } Q_d < \tau_3 \end{cases} \qquad (9.4)$$

where $Q_d$ is the estimated quality score based on dark pixels and it can be computed by a method discussed in Section 8.1.2 using a single threshold, i.e., $I_c < T_L$.

ALE technique considers four thresholds to label the quality of images. For example, $Q_d \geq \tau_1$ represents normal images, $Q_d \geq \tau_2$ and $Q_d < \tau_1$ indicate shadow images, $Q_d \geq \tau_3$ and $Q_d < \tau_2$ represent dark images and $Q_d < \tau_3$ shows very dark images. The value of $e$ is assigned according to the quality of images for adaptive logarithm enhancement using Equation 9.4.

(a)

(b)  $e = 1$

(c)  $e = 0.1$

(d)  $e = 0.01$

(e)  $e = 0.001$

**Figure 9.2.**   Obtained different enhanced images for different values of $e$ [13].

### 9.2.2 *Experimental results*

The technique ALE proposed in Ref. [13] has been evaluated on 100 images of different sizes ($720 \times 480$ to $3264 \times 2448$) of the Wang database [15]. The technique is compared with AHE [16], Zhang [17], MSRCR [18], AINDANE [19], Swang [15] and SGNE [20]. The ALE results are compared with other techniques in two ways, *viz.* subjective assessment and objective assessment. First, we make a subjective comparison, which shows the robustness of the ALE in terms of image enhancement and naturalness preservation. Further, we objectively evaluate the performance using LOE [15]. We have considered two images in subjective assessment and enhanced them with various existing techniques. Figures 9.3 and 9.4 present results for these images obtained after using various image enhancement techniques. Figure 9.3 shows results for a very dark image captured during the daytime, whereas Figure 9.4 shows results for an image captured at nighttime with some bright spots.

From Figure 9.3, it is seen that the techniques AHE [16], AINDANE [19] and MSRCR [18] have failed to enhance this very dark image. Techniques like Zhang [17] and SGNE [20] have over-enhanced the original image and have resulted in some additional bright and dark spots in the enhanced images. The technique in Swang [15] performs well and produces good enhancement; however, the pleasing perceptual quality achieved in the enhanced image is not as appealing as obtained by the ALE. The ALE technique not only effectively enhances the quality of the original image but also preserves the naturalness of the image to get pleasing perceptual quality. Figure 9.4 elucidates that techniques such as AHE [16], AINDANE [19] and Swang [15] have over-enhanced the original image. In the MSRCR [18] technique, it has not much improved the original image. Techniques Zhang [17] and SGNE [20] have performed relatively better; however, some artifacts can be observed in the cloud area. The ALE technique has effectively enhanced the original image and produced no artifacts.

We have included LOE for objective evaluation of the ALE technique. A smaller value of LOE shows better preservation of lightness order and naturally pleasing appearance. The average LOE values for these techniques have been presented in Table 9.2. We observe from Table 9.2 that the ALE has obtained the lowest average LOE value as compared to all other techniques. This, in turn, shows that the ALE preserves lightness order better as compared to other existing techniques.

**Figure 9.3.** Demonstration of image enhancement results: (a) Original image and different results of image enhancement obtained by techniques of (b) AHE [16], (c) AIN-DANE [19], (d) MSRCR [18], (e) Swang [15], (f) Zhang [17], (g) SGNE [20] and (h) ALE techniques.

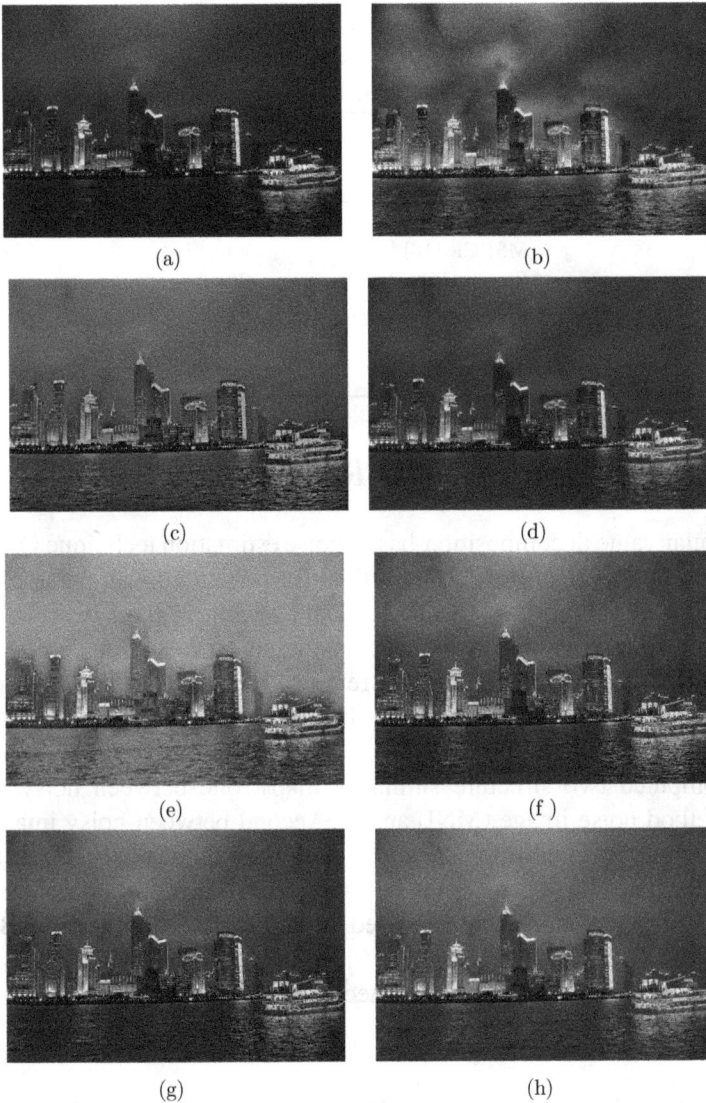

**Figure 9.4.** Demonstration of image enhancement results: (a) Original image and different results of image enhancement obtained by techniques of (b) AHE [16], (c) AINDANE [19], (d) MSRCR [18], (e) Swang [15], (f) Zhang [17], (g) SGNE [20] and (h) ALE techniques.

**Table 9.2.** Average LOE value of different techniques for illumination enhancement.

| Techniques | LOE |
| --- | --- |
| AHE [16] | 462.87 |
| AINDANE [19] | 729.53 |
| MSRCR [18] | 480.17 |
| Swang [15] | 291.44 |
| Zhang [17] | 448.59 |
| SGNE [20] | 244.77 |
| **ALE** | **203.96** |

# 9.3 Quality-Aware Noise Reduction

A singular value decomposition-based noise estimation technique Q-metric has been proposed in Ref. [21]. Further, for noise removal, parameters of two denoising algorithms BM3D [22] and SKR [23] have been set automatically to remove noise in images using an estimated quality score of an image. Another automatic noise removal technique has been proposed in Ref. [24]. This technique has searched optimal parameters for a denoising method by estimating the quality of the noisy images. The technique has computed two structure similarity maps: one between noisy image and method noise image (MNI) and the second between noisy image and denoised image. Finally, the linear correlation coefficient between structure similarity maps has been computed as a quality score for an input image. This estimated quality score is used to automate the denoising BM3D method using the method suggested in Ref. [21]. A quality-aware adaptive Wiener filter-based noise reduction technique is presented in the following subsection.

## 9.3.1 *Quality-aware Wiener filter (QAWF)*

The existing techniques in the literature often apply the enhancement technique blindly on images of different kinds without considering the nature of images and the level of distortions present in them. For example, to enhance images of a database, a Wiener filter of fixed size is used on all the images to make them noise-free. However, a filter of the same size may not reduce the noise properly, as the level of noise present in different images may vary. Hence, there is a need to choose the size of the filter adaptively based on the level of distortion present in the image. In this section, we propose a quality-aware Wiener filter for denoising of images. For designing an adaptive filter to denoise an image, we first estimate the quality of the given noisy image and then choose an appropriate filter size based on the obtained quality score.

### 9.3.1.1 *Computation of filter size*

Let the size of the Wiener filter be ($[F_s \times F_s]$). We compute the size based on the estimated quality score of the image and give the value of $F_s$ as follows:

$$F_s = C \times Q_N \tag{9.5}$$

where $Q_N$ is the obtained quality score of the image estimated using the noise information by technique proposed in Ref. [25] and $C$ is a constant whose value is chosen empirically.

Before applying the QAWF denoising technique, the quality of an input image is analyzed to know if the image is noisy or not. If the quality of an input image is good (i.e., $Q_N \geq \eta_N$), then there is no need to apply the proposed denoising technique. The image is considered noise-free if $Q_N \geq \eta_N$, where $\eta_N$ is a threshold obtained experimentally. Image quality below this threshold is considered poor, and then we apply a quality-aware Wiener filter technique for denoising.

To see the results visually, we show a few noisy images and their corresponding denoised images computed by the quality-aware adaptive Wiener filter in Figure 9.5. It is seen from the obtained enhancement results that the adaptive Wiener filter has performed the denoising task very well.

### 9.3.1.2 *Experimental results*

Experimental analysis of the QAWF is performed on LIVE [26], CSIQ [27] and TID08 [28] databases. For the objective analysis, PSNR between an original image and the enhanced image is computed where a higher value of PSNR indicates better enhancement.

Results are analyzed in three different cases. In the first case, we analyze noisy images available before noise reduction. In the second case, the analysis of the images obtained by performing denoising using a traditional Wiener filter is considered. In the third case, images were obtained

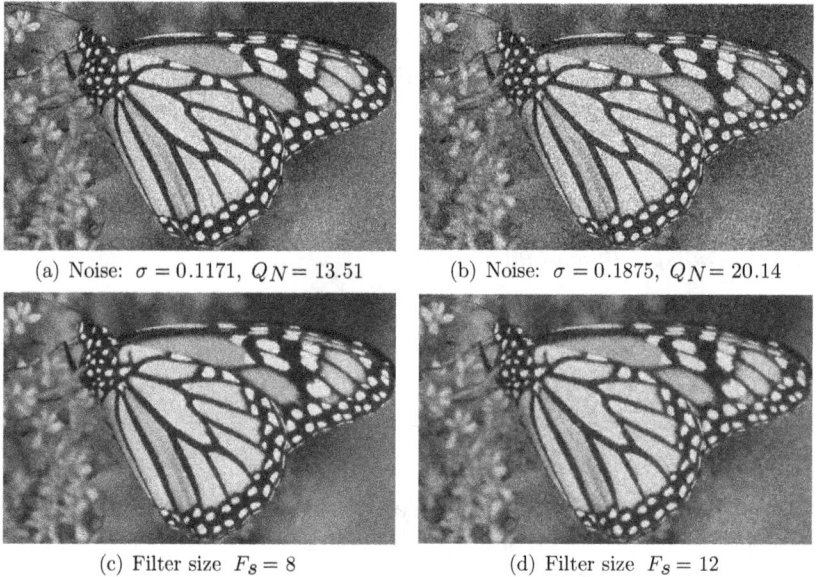

(a) Noise: $\sigma = 0.1171$, $Q_N = 13.51$     (b) Noise: $\sigma = 0.1875$, $Q_N = 20.14$

(c) Filter size $F_s = 8$     (d) Filter size $F_s = 12$

**Figure 9.5.** Few noisy images (first two) with their corresponding denoised versions (last two) [25].

by performing denoising using the QAWF analyzed. In the first case, PSNR value is computed between the original image and the corresponding degraded noisy image. In the second case, a Wiener filter of fixed size is applied to all the images of the database and the PSNR value is obtained between the original image and the enhanced image. In the third case, the size of Wiener filter is adaptively chosen for each individual image based on its quality score to perform denoising. Further, PSNR value is computed between the original image and the enhanced image obtained by applying the adaptive Wiener filter.

Figure 9.6 presents an experimental analysis on LIVE database, which consists of 174 noisy images. It shows the total PSNR value of all 174 images obtained for various fixed-sized Wiener filters and the QAWF technique. Figure 9.7 presents results in terms of PSNR value on CSIQ database, which consists of 150 noisy images, whereas Figure 9.8 shows results on TID08 database, having 100 noisy images. Figures 9.6, 9.7 and 9.8 show the total PSNR value of all the images of the respective databases corresponding to Wiener filters of fixed sizes as well as for the QAWF technique. It is evident from the figures that the denoised images obtained by the QAWF technique (Point A in Figures 9.6, 9.7 and 9.8) produce higher

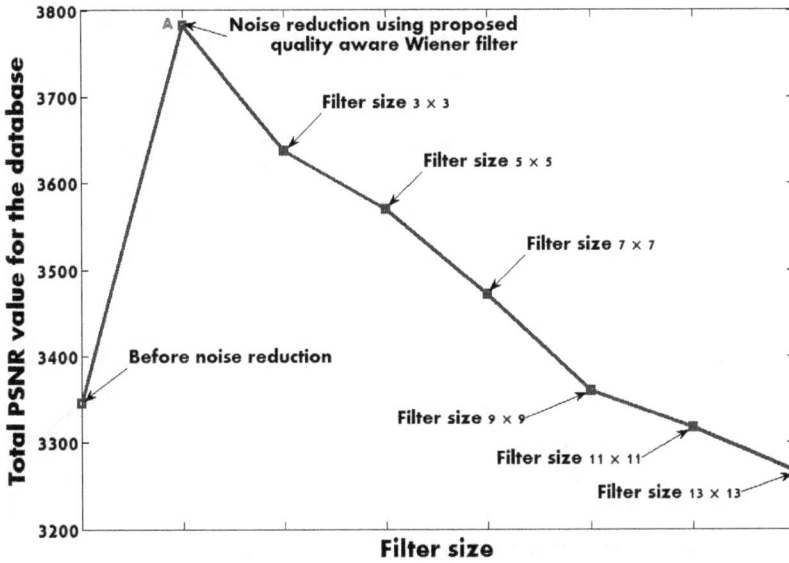

**Figure 9.6.** Total PSNR value of the images of LIVE database.

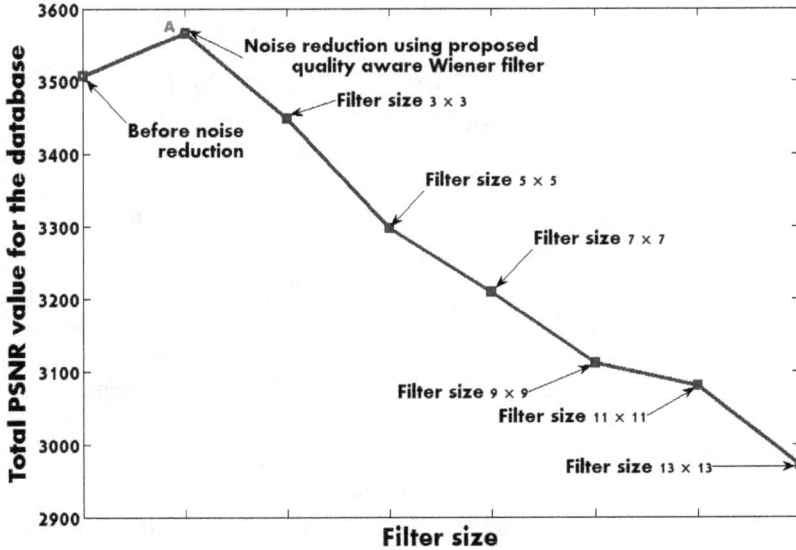

**Figure 9.7.** Total PSNR value of the images of TID08 database.

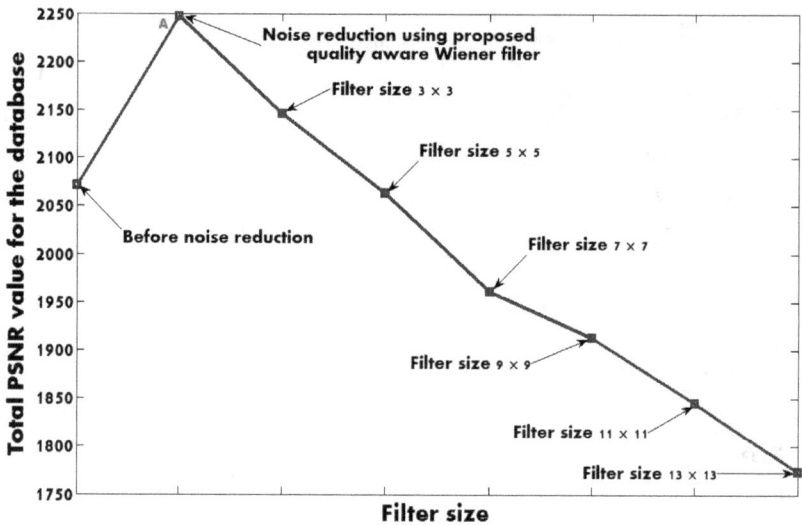

**Figure 9.8.**   Total PSNR value of the images of CSIQ database.

total PSNR values for all the databases as compared to the denoised images obtained for fixed-sized Wiener filters.

## 9.3.2   *Quality-aware noise reduction using multiple denoising algorithms*

The reduction of noise in images is a fundamental challenge in computer vision and image processing, with wide-ranging applications from medical imaging to surveillance systems. Noise in digital images degrades their quality and obstructs subsequent analysis and interpretation. Conventional denoising methods typically utilize fixed algorithms, which may not yield optimal results across various image datasets. In our progressively data-driven world, the demand for adaptive image quality enhancement methods is crucial. The prevalence of noisy images in real-world situations requires inventive strategies for image denoising. Tackling this challenge holds the promise of improving the quality and dependability of image-driven decision-making systems, such as autonomous vehicles, medical diagnosis and facial recognition. The importance of this research is underscored by the growing need for enhanced image-denoising methods.

   In this chapter, we present a Quality Aware Adaptive Denoising (QAAD) approach that utilizes the Blind/Referenceless Image Spatial

Quality Evaluator (BRISQUE) [29] as a crucial criterion for choosing the most appropriate denoising algorithm for a given image. Our main goals include evaluating image quality through BRISQUE scores, creating a mechanism for selecting a denoising algorithm specific to the image's characteristics and showcasing the efficacy of the presented approach in enhancing image quality. In the process of determining the most appropriate image scoring algorithm for our dataset, we initially applied BRISQUE, NIQE [30] and BLIINDS [31] (Blind Image Integrity Notator using DCT and SVM). The objective was to identify which of these algorithms effectively evaluated image quality. Following thorough evaluation, BRISQUE emerged as the preferred choice. BRISQUE demonstrated the capacity to provide a significant range of quality scores, establishing it as a robust metric for our purposes. Additionally, BRISQUE employed natural scene statistics to assess image quality, aligning well with our objectives. Moreover, BRISQUE consistently showcased its ability to reliably quantify improvements in image quality, distinguishing itself from other scoring algorithms and reinforcing our decision to adopt it as our primary quality assessment metric.

Expanding on our preliminary discovery that BRISQUE is the optimal image-scoring algorithm, we proceeded to formulate our adaptive denoising approach. Alongside the LIVE Multi-Distortion Image Database, we integrated three standard datasets: CSIQ [32] (Categorical Image Quality Database), SIDD [33] (SIDD Image Denoising Dataset) and TID2013 [28] (TID Image Database 2013). This extension enabled us to comprehensively evaluate the adaptability of QAAD methodology across a broad spectrum of real-world scenarios. Utilizing these expanded datasets, we subjected them to a comprehensive set of image denoising algorithms, including Wiener [34], Wavelet [34], Gaussian [34], Median [34], NAFNet [35] and MIRNet [36]. Through a systematic evaluation of the performance of these algorithms in conjunction with BRISQUE scores, we gained insights into identifying the optimal algorithm for specific quality ranges. This methodology enabled the development of an adaptive denoising framework that customizes the algorithm selection based on the inherent noise characteristics of each image, ensuring superior denoising results.

QAAD methodology integrates dataset expansion, thorough assessment of image denoising algorithms and the consistent use of BRISQUE for quality evaluation. These modifications guarantee the robustness and efficacy of an adaptive image-denoising approach, laying the foundation for improved image quality across a variety of real-world scenarios. The remaining sections of this chapter are structured as follows: Section 9.3.2.1 presents a detailed exploration of the algorithms utilized for image denoising, providing an in-depth understanding of the computational techniques central to our study. Section 9.3.2.2 outlines the sequential pipeline that guides an input image through this methodology, resulting in a denoised output. In Section 9.3.2.3, we elaborate on the experimental procedures conducted on the dataset, supplemented by an example. Section 9.3.2.4 performs a meticulous analysis of the obtained results, offering insights into the performance of the presented methodology. Finally, Section 9.3.2.5 concludes this chapter by summarizing findings, discussing implications and suggesting avenues for future research, encapsulating the essential contributions of this study within an academic context.

### 9.3.2.1 *Preliminaries*

Within this section, we furnish thorough explanations of the core algorithms that underpin our research. These algorithms play a crucial role in the methodologies and solutions presented in this study. BRISQUE serves as our primary image quality metric, while the Wiener filter, median filter, Gaussian filter, wavelet transform, NAFNet and MIRNet are employed for image denoising. These algorithms constitute the foundation of this chapter, laying the groundwork for the subsequent insights and discoveries.

#### 9.3.2.1.1 BRISQUE

The Blind/Referenceless Image Spatial Quality Evaluator (BRISQUE) algorithm is a no-reference image quality assessment method designed to evaluate the perceptual quality of images without the need for a reference image. BRISQUE functions by extracting statistical features from the image in the spatial domain, such as the mean and standard deviation of image patches. These features are then utilized to train a Support Vector Machine (SVM) model, which can predict the image quality based on learned patterns from a substantial dataset of images. A lower BRISQUE score serves as a key indicator of superior image quality, playing a crucial role in interpreting the results presented in this research. Figure 9.9 presents the score computed by the BRISQUE algorithm for original, noisy and blurry images.

Applying the BRISQUE algorithm to an image involves these four concise steps:

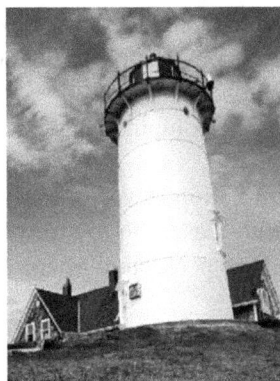

(a) Original Image.
BRISQUE score = 20.65

(b) Noisy Image.
BRISQUE score = 52.60

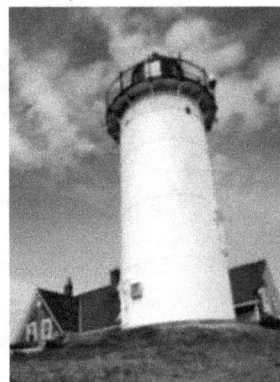

**Figure 9.9.** Quality score computed by BRISQUE for different images [29].

1. **Patch Extraction:** Divide the image into non-overlapping patches.
2. **Statistical Features:** Compute statistical features (e.g., mean and standard deviation) for each patch.
3. **Feature Vector:** Aggregate computed features into a vector to represent spatial characteristics.
4. **SVM Prediction:** Train a support vector machine on a dataset, and use it to predict image quality based on the feature vector. This allows for no-reference assessment of perceptual image quality.

### 9.3.2.1.2   Wiener filter

The Wiener filter, a classical image denoising technique, endeavors to recover an image from noise by estimating the power spectral density (PSD) of the noise. In this section, we outline the methodology and mathematical formulation of the Wiener filter, accompanied by a sample image result and a performance graph applied to the entire dataset. Figure 9.10 illustrates an instance of denoising using a Wiener filter.

The Wiener filter follows a systematic approach to denoising:

1. Obtain a noisy image and the estimated power spectral density of the noise.
2. Compute the Fourier transform of the noisy image and the PSD of the noise.
3. Apply the Wiener filter formula in the frequency domain to obtain the filtered image's Fourier transform.
4. Inverse Fourier transforms the filtered image to obtain the denoised image in the spatial domain.

Original Image                    Denoised Image

**Figure 9.10.**   An example of Wiener filtering [37].

In the frequency domain, the Wiener filter applies a deconvolution operation based on the estimated PSD of the noise $(N(f))$ and the Fourier transform of the noisy image $(Y(f))$. The filtered image's Fourier transform, denoted as $X(f)$, is calculated using the following formula:

$$X(f) = \frac{H(f)}{H(f) + \frac{1}{S(f)}} \tag{9.6}$$

where $H(f)$ is the Fourier transform of the noisy image. $S(f)$ is the estimated power spectral density of the noise.

### 9.3.2.1.3 Median filter

The median filter, a widely employed image denoising technique, entails replacing each pixel's value with the median value within a specified window. This section provides an overview of the methodology and procedure for applying the median filter, accompanied by a sample image result and a performance graph. Figure 9.11 illustrates an instance of denoising using a median filter.

The median filter follows these steps for image denoising:

1. Define the filter window size (in our case, 5 x 5).
2. Traverse the entire image.
3. For each pixel, find the median value within the window.
4. Assign the median value to that particular pixel.
5. Repeat this process for all pixels in the image.

### 9.3.2.1.4 Gaussian filter

The Gaussian filter is a commonly utilized image processing technique for image smoothing. It diminishes noise and details in an image by applying a Gaussian kernel to each pixel. This section outlines the methodology and process of applying the Gaussian filter, accompanied by a sample image result. Figure 9.12 illustrates an example of denoising using a Gaussian filter.

The process of Gaussian filtering is as follows:

1. Define the Gaussian kernel size (kernel_size) and standard deviation $(\sigma)$. In our case, $\sigma = 0.6$.
2. Generate a 2D Gaussian kernel of size kernel_size $\times$ kernel_size using the chosen $\sigma$. This kernel represents the Gaussian distribution.

(a) Original Image

(b) Denoised Image

**Figure 9.11.** An example of median filtering [38].

3. Iterate through each pixel in the image, excluding the border pixels where the kernel may not fit entirely.
4. For each pixel, place the center of the Gaussian kernel on the pixel.
5. Multiply the pixel values within the kernel by the corresponding values in the Gaussian kernel.
6. Sum up the results to calculate the new pixel value.
7. Set the pixel's value to the computed sum.

### 9.3.2.1.5 Wavelet transform

The Wavelet transform is a potent technique for both image denoising and feature extraction. It dissects the image into various scales of detail

(a) Original Image

(b) Denoised Image

**Figure 9.12.** An example of Gaussian filtering [38].

and approximation using a selected wavelet transform, such as Discrete Wavelet Transform (DWT) or Continuous Wavelet Transform (CWT). Figure 9.13 provides an illustration of denoising through a Wavelet filter. The steps in applying the Wavelet filter are as follows:

1. Apply the chosen wavelet transform (DWT or CWT) to decompose the image into multiple scales of detail and approximation.

(a) Original Image

(b) Denoised Image

**Figure 9.13.** An example of wavelet filtering [38].

2. Apply thresholding to the coefficients at each scale to reduce noise. The thresholding method can be hard or soft, depending on the application. In our experiment, we have used BayesShrink algorithm [39] which is an adaptive approach to wavelet soft thresholding where a unique threshold is estimated for each wavelet sub-band.

3. Reconstruct the denoised image using the inverse wavelet transform, combining the modified coefficients from all scales.

(a) Original Image

(b) Denoised Image

**Figure 9.14.** An example of NAFNet filtering [38].

### 9.3.2.1.6 NAFNet denoising

The NAFNet (Nonlinear Activation Free Network) filter represents an advanced image denoising technique that utilizes a U-shaped neural network. This network is specifically crafted to handle noisy images and generate clean, denoised counterparts. Figure 9.14 provides an example of denoising with an NAFNet filter. The steps involved in the NAFNet process are as follows:

1. **U-shaped Network**: The noisy image is passed through a U-shaped network. This network comprises two main paths: an encoding path and

a decoding path. The encoding path extracts features from the noisy image, while the decoding path reconstructs the clean image from these features.

2. **Training with Policy Gradient**: The NAFNet is trained using a policy gradient method. The policy gradient method is a reinforcement learning technique that updates the network's parameters based on the rewards it receives during training. These rewards are assigned based on the similarity between the reconstructed image and the ground truth (clean) image. This training method helps the network learn to denoise images effectively.

3. **Efficiency through Nonlinear Activation Functions**: Notably, the NAFNet architecture may not require the use of nonlinear activation functions. By reducing the use of nonlinearities, the network becomes more efficient and easier to train. This is a unique feature of the NAFNet approach that sets it apart from traditional neural network denoising methods.

#### 9.3.2.1.7   MIRNet denoising

The MIRNet algorithm, initially developed for single-image super-resolution, features a unique architecture tailored for comprehensive feature extraction across multiple scales. Its notable attributes include a multi-scale feature extraction model, facilitating the computation of complementary features while preserving original high-resolution details. Additionally, MIRNet adopts a recursive residual design, progressively breaking down input signals to simplify the learning process and enable the construction of deep networks. While MIRNet is traditionally utilized for single-image super-resolution, its application for image denoising underscores its adaptability and effectiveness across various image enhancement tasks. This versatility is supported by MIRNet's inherent ability to capture intricate spatial details and features, making it well suited for denoising applications where preserving fine details is crucial for optimal results. Figure 9.15 provides an example of denoising with an MIRNet filter. The steps involved in applying the MIRNet algorithm are as follows:

1. Extract features from the input image across multiple spatial scales using the MIRNet's feature extraction model. Ensure the preservation of high-resolution details during this process.

(a) Original Image

(b) Denoised Image

**Figure 9.15.** An example of MIRNet filtering [38].

2. Implement a recurrent mechanism for information exchange, gradually fusing features obtained from multi-resolution branches. This step enhances representation learning by combining contextual information from different scales.

3. Utilize a selective kernel network to dynamically combine variable receptive fields, facilitating the fusion of multi-scale features. Ensure that the original feature information is faithfully preserved at each spatial resolution.

4. Implement the recursive residual design of MIRNet, breaking down the input signal progressively. This design simplifies the learning process and enables the construction of deep networks, enhancing the algorithm's ability to capture intricate spatial details.

### 9.3.2.2 *Pipeline of quality-aware adaptive denoising (QAAD)*

The QAAD workflow is depicted in Figure 9.16. Within the established framework, where user-supplied images act as the primary input, the process begins with the utilization of the BRISQUE image-scoring algorithm. This algorithm effectively assesses the perceptual quality of each provided image, assigning an initial quality score that serves as a critical metric for subsequent enhancements. Upon determining the BRISQUE scores, the system dynamically selects an appropriate denoising algorithm based on the derived quality assessments. This adaptive approach ensures that the chosen denoising technique is customized to the specific characteristics of each image, optimizing the overall improvement in image quality.

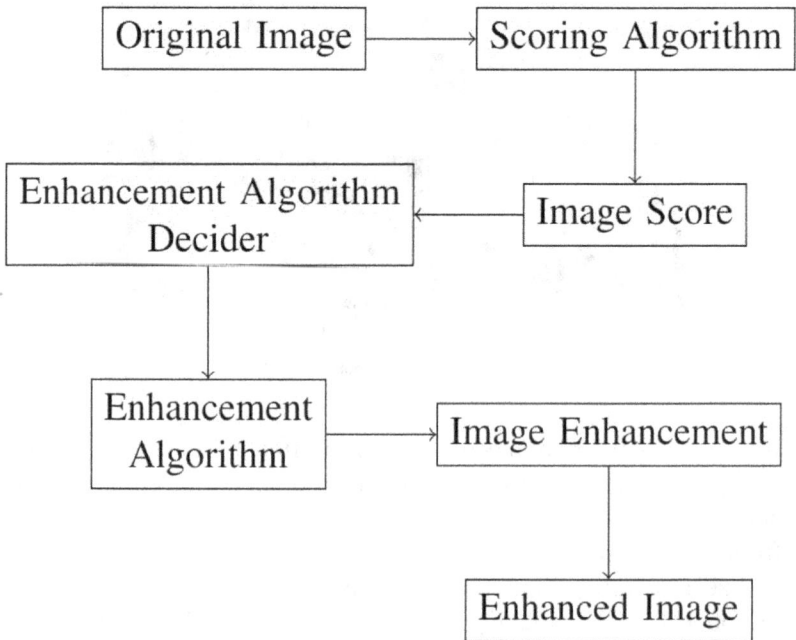

**Figure 9.16.** The pipeline for quality-aware adaptive denoising.

Additionally, the chosen denoising algorithm is applied to the input image, employing advanced techniques to diminish noise and artifacts while preserving essential details. This step is crucial in refining the visual aesthetics of the image and alleviating any distortions or imperfections present in the original input. The refined image serves as a demonstration of the effectiveness of the selected denoising algorithm. This showcase not only highlights the system's capability to enhance image quality but also underscores the significance of personalized and adaptive algorithms in achieving optimal results across a diverse range of input images. In summary, the integration of the BRISQUE image-scoring algorithm and adaptive denoising techniques in this framework ensures a sophisticated and user-centric approach to image enhancement. It addresses the unique characteristics and quality nuances of each user-provided image.

### 9.3.2.3  *Experiment*

In our experimental framework, we employed a combination of image processing algorithms, incorporating the BRISQUE image-scoring algorithm, as well as various denoising techniques such as the Wiener filter, wavelet transform, median filter, Gaussian denoising, NAFNet deep learning and MIRNet denoising. This comprehensive suite serves as the foundational apparatus for this image quality enhancement framework. Utilizing the BRISQUE image-scoring algorithm, we compute the baseline quality score for each input image. Following this, a denoising algorithm is dynamically chosen based on this BRISQUE score, resulting in an effective enhancement of image quality. The enhanced image serves as a demonstration of the efficacy of the selected denoising algorithm.

Figure 9.17 presents the comprehensive workflow of the experiment. Initially, the QAAD methodology integrates a meticulous denoising algorithm selection process using four benchmark datasets: LIVE, CSIQ, TID13 and SIDD. To commence this process, we subject the original images within each dataset to BRISQUE scoring. The obtained scores are then utilized to categorize the images into distinct groups based on their original BRISQUE scores, such as [1,6), [6,11), [11,16), and so forth.

To provide a specific example, let's focus on the [31, 36) score group. Within this range, there are 10 images in CSIQ, 8 in LIVE, 30 in SIDD and 56 in TID13, each with an original BRISQUE score. By applying all six denoising algorithms to each image across all datasets, we calculate the post-denoising BRISQUE scores. Therefore, for each dataset and each

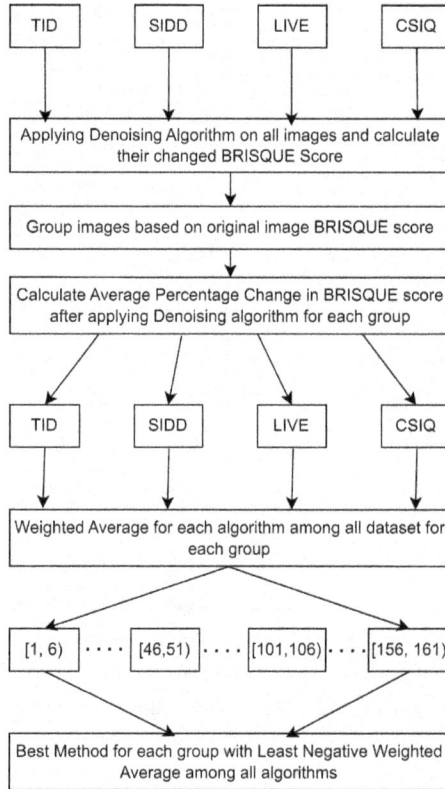

**Figure 9.17.**    Experiment Workflow.

image, we obtain the original score and six subsequent scores corresponding to the application of the denoising algorithms. Additionally, we compute the group average for each denoising algorithm within the [31, 36) score range for each dataset. This results in six averages for the six algorithms, regardless of the number of images in the group for each dataset.

Repeat this procedure for all six denoising algorithms within the [31, 36) score group, yielding a set of six weighted averages. Ultimately, we determine the best-performing denoising algorithm for the [31, 36) score range by selecting the algorithm with the lowest weighted average (lower BRISQUE implies better quality). This illustrates the thorough evaluation process conducted across different score ranges. We choose to implement the algorithm only if the lowest weighted average is negative, as a decrease in BRISQUE score signifies an improvement in quality.

## 9.3.2.4 *Results Analysis*

In a thorough examination of denoising methods applied to images across various BRISQUE value ranges, distinct patterns and trends have surfaced. The BRISQUE value ranges were discretized into groups, each representing a specific range of image quality scores, while denoising methods were assigned to each group based on their respective efficacy. The results obtained reveal a nuanced strategy in optimizing denoising algorithms according to the inherent quality characteristics of the images. Denoising methods were assigned to each group based on the lowest weighted negative average of that group. Table 9.3 presents the experiment outcome, comprising three columns: "Low", "High" and "Best Method." The "Low" and "High" columns define score ranges [Low, High) for each row. When an image's BRISQUE score falls within a specified range, the corresponding denoising "Best Method" is applied. For example, if the BRISQUE score lies between 21 and 36 (exclusive), the Wiener denoising method is employed, showcasing the adaptive selection of denoising techniques based on image quality assessments. It is crucial to note that, due to the absence of images from any dataset in the original BRISQUE score range [141, 146) and [151, 156), we pragmatically extended the successful denoising method, MIRNet, from the adjacent range [136, 161). This decision was based on the observed effectiveness of MIRNet in the neighboring [136, 141), [146, 151) and [156, 161) ranges.

The results presented in Table 9.4 illustrate the average change in BRISQUE scores for various denoising algorithms, including Gaussian, median, wavelet, Wiener, MIRNet, NAFNet and presented framework (QAAD), across different datasets. In comparison to individual algorithms, the QAAD consistently exhibits superior performance, highlighting its

**Table 9.3.** Experiment Outcome.

| Low ($\geq$) | High ($<$) | Best Method |
|---|---|---|
| 1 | 21 | None |
| 21 | 36 | Wiener |
| 36 | 106 | Gaussian |
| 106 | 111 | NAFNet |
| 111 | 121 | MIRNet |
| 121 | 136 | Gaussian |
| 136 | 161 | MIRNet |

**Table 9.4.**    Improvement in BRISQUE Score after denoising.

| | | Gaussian | Median | Wavelet | Wiener | MIRNet | NAFNet | QAAD |
|---|---|---|---|---|---|---|---|---|
| | | | | | Denoising Algorithm | | | |
| Dataset | TID13 | 0.93 | 97.77 | −16.59 | −11.62 | 42.81 | 49.92 | −33.39 |
| | CSIQ | 52.17 | 174.59 | −5.17 | 2.44 | 89.83 | 86.30 | −17.69 |
| | SIDD | 27.54 | 47.03 | 15.85 | −30.57397 | - | - | −20.57 |
| | LIVE | −4.36 | 21.21 | 17.74 | −8.00 | 99.26 | 113.07 | −33.48 |

adaptability in selectively applying algorithms to enhance image quality. Figure 9.18 depicts the percentage change in BRISQUE scores after applying the top four denoising algorithms on the LIVE dataset. As observed in the graphs, QAAD surpasses other denoising algorithms when applied alone to the dataset. Specifically, the presented method yields significant improvements, as indicated by negative values, underscoring its effectiveness in diverse scenarios. This adaptive approach underscores the potential of QAAD framework to outperform standalone denoising methods across a range of datasets.

### 9.3.2.5    *Conclusions and future work*

The QAAD method, employing dynamic denoising algorithm selection based on BRISQUE scores, surpasses individual algorithms across diverse datasets, demonstrating its adaptability and effectiveness in improving image quality. Looking ahead, it is essential to address fixed hyperparameters in algorithms, such as median, Gaussian and Wiener. Future work could concentrate on making these hyperparameters adaptive or providing a range of options to enhance flexibility. Expanding the denoising algorithm repertoire is another potential avenue. While the current framework includes six algorithms like median, Gaussian and MIRNet, further research could explore the integration of additional algorithms to broaden the set of denoising options. An exciting prospect involves the integration of deep learning, particularly Convolutional Neural Networks (CNNs), to directly learn image quality and dynamically select denoising algorithms. This holds the potential for a more nuanced and adaptive solution beyond traditional categorization based on BRISQUE scores. In essence, these future directions aim to enhance the adaptability, flexibility and overall

## Weiner Enhancement Technique

## Gaussian Enhancement Technique

## Wavelet Enhancement Technique

## Proposed Framework Denoising

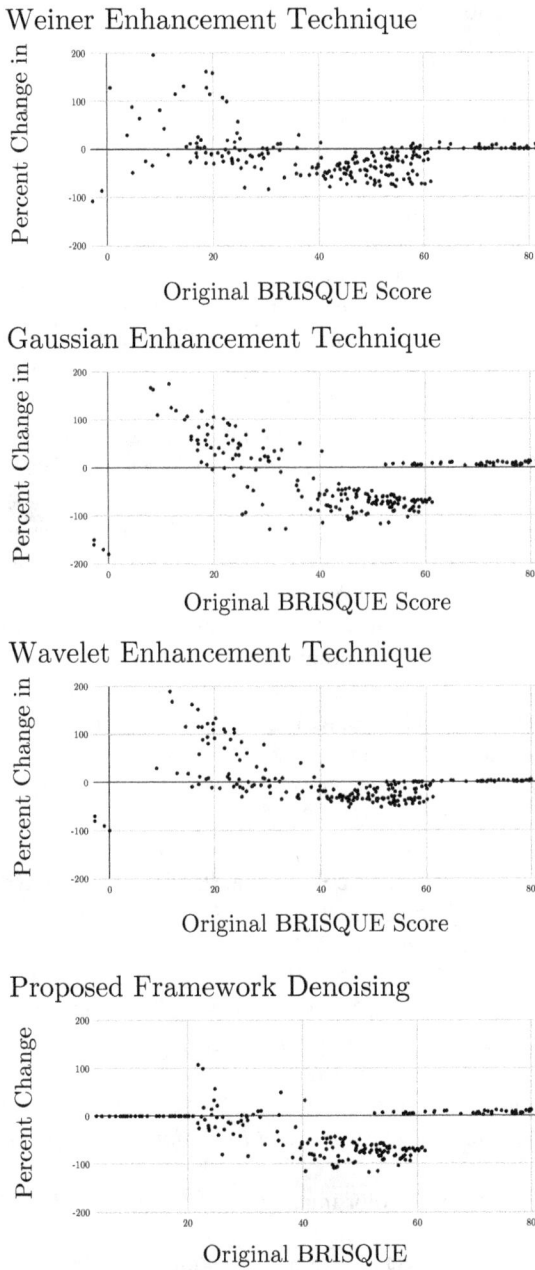

**Figure 9.18.** Change in BRISQUE score on the LIVE Dataset.

efficacy of the denoising framework, driving advancements in image processing and quality improvement.

# References

[1]  A. Saha and Q. Wu, "Utilizing image scales towards totally training free blind image quality assessment," *IEEE Transactions on Image Processing*, 24(6), 1879–1892, 2015.

[2]  P. Joshi and S. Prakash, "A quality aware technique for biometric recognition," in *Proceedings of International Conference on Signal Processing and Integrated Networks (SPIN 2015)*, 2015, pp. 795–800.

[3]  B. Subramanyam, P. Joshi, M. K. Meena, and S. Prakash, "Quality based classification of images for illumination invariant face recognition," in *Proceedings of IEEE International Conference on Identity, Security and Behavior Analysis (ISBA)*, 2016, pp. 1–6.

[4]  D. Karaboga and B. Basturk, "A powerful and efficient algorithm for numerical function optimization: Artificial bee colony (ABC) algorithm," *Journal of Global Optimization*, 39(3), 459–471, 2007.

[5]  K. Gu, S. Wang, G. Zhai, W. Lin, X. Yang, and W. Zhang, "Analysis of distortion distribution for pooling in image quality prediction," *IEEE Transactions on Broadcasting*, 62(2), 446–456, 2016.

[6]  K. Shaheed, L. Yang, G. Yang, I. Qureshi, and Y. Yin, "Novel image quality assessment and enhancement techniques for finger vein recognition," in *International Conference on Security, Pattern Analysis, and Cybernetics (SPAC)*, 2018, pp. 223–231.

[7]  P. Joshi and S. Prakash, "An efficient technique for image contrast enhancement using artificial bee colony," in *IEEE International Conference on Identity, Security and Behavior Analysis (ISBA 2015)*, 2015, pp. 1–6.

[8]  D. B. Fogel, *Evolutionary Algorithms in Theory and Practice*. Wiley Online Library, Hoboken, NJ, 1997.

[9]  S. Hashemi, S. Kiani, N. Noroozi, and M. E. Moghaddam, "An image contrast enhancement method based on genetic algorithm," *Pattern Recognition Letters*, 31(13), 1816–1824, 2010.

[10]  A. Draa and A. Bouaziz, "An artificial bee colony algorithm for image contrast enhancement," *Swarm and Evolutionary Computation*, 16, 69–84, 2014.

[11]  "Kodak lossless true color image suite." Available: http://r0k.us/graphics/kodak/.

[12]  K. Lee, J. Ho, and D. Kriegman, "Acquiring linear subspaces for face recognition under variable lighting," *IEEE Transactions on PAMI*, 27(5), 684–698, 2005.

[13] P. Joshi and S. Prakash, "Image enhancement with naturalness preservation," *Visual Computer*, 36(1), 71–83, 2020.

[14] C. Wu, S. Tulyakov, and V. Govindaraju, "Image quality measures for fingerprint image enhancement," in *Multimedia Content Representation, Classification and Security*, 2006, pp. 215–222.

[15] S. Wang, J. Zheng, H. M. Hu, and B. Li, "Naturalness preserved enhancement algorithm for non-uniform illumination images," *IEEE Transactions on Image Processing*, 22(9), 3538–3548, 2013.

[16] S. M. Pizer, E. P. Amburn, J. D. Austin, R. Cromartie, A. Geselowitz, T. Greer, B. T. H. Romeny, and J. B. Zimmerman, "Adaptive histogram equalization and its variations," *Computer Vision, Graphics, and Image Processing*, 39(3), 355–368, 1987.

[17] H. Zhang, Y. Li, H. Chen, D. Yuan, and M. Sun, "Perceptual contrast enhancement with dynamic range adjustment," *Optik – International Journal for Light and Electron Optics*, 124(23), 5906–5913, 2013.

[18] D. J. Jobson, Z. Rahman, and G. A. Woodell, "A multiscale retinex for bridging the gap between color images and the human observation of scenes," *IEEE Transactions on Image Processing*, 6(7), 965–976, 1997.

[19] L. Tao and V. K. Asari, "Adaptive and integrated neighborhood-dependent approach for nonlinear enhancement of color images," *Journal of Electronic Imaging*, 14(4), 043006 (1–14), 2005.

[20] Y. Li, H. Zhang, W. Jia, D. Yuan, F. Cheng, R. Jia, L. Li, and M. Sun, "Saliency guided naturalness enhancement in color images," *Optik - International Journal for Light and Electron Optics*, 127(3), 1326–1334, 2016.

[21] X. Zhu and P. Milanfar, "Automatic parameter selection for denoising algorithms using a no-reference measure of image content," *IEEE Transactions on Image Processing*, 19(12), 3116–3132, 2010.

[22] K. Dabov, A. Foi, V. Katkovnik, and K. Egiazarian, "Image denoising by sparse 3-d transform-domain collaborative filtering," *IEEE Transactions on Image Processing*, 16(8), 2080–2095, 2007.

[23] H. Takeda, S. Farsiu, and P. Milanfar, "Kernel regression for image processing and reconstruction," *IEEE Transactions on Image Processing*, 16(2), 349–366, 2007.

[24] X. Kong, K. Li, Q. Yang, L. Wenyin, and M.-H. Yang, "A new image quality metric for image auto-denoising," in *Proceedings of IEEE International Conference on Computer Vision*, 2013, pp. 2888–2895.

[25] P. Joshi and S. Prakash, "Nr-iqa for noise-affected images using singular value decomposition," *IET Signal Processing*, 13(2), 183–191, 2019.

[26] L. C. H.R. Sheikh, Z. Wang and A. Bovik, "LIVE image quality assessment database release 2, LIVE, University of Texas at Austin, 2006. Available: http://live.ece.utexas.edu/research/quality."

[27] E. C. Larson and D. M. Chandler, "Most apparent distortion: full-reference image quality assessment and the role of strategy," *Journal of Electronic Imaging*, 19(1), 011 006(1–21), 2010.

[28] A. Z. N. Ponomarenko, V. Lukin, K. Eziazarian, M. Carli, and F. Battisti, "TID2008— a database for evaluation of full reference image quality metrics," *Advance Modern Radioelectronics*, 10(4), 30–45, 2009.

[29] A. Mittal, A. K. Moorthy, and A. C. Bovik, "No-reference image quality assessment in the spatial domain," *IEEE Transactions on Image Processing*, 21(12), 4695–4708, 2012.

[30] A. Mittal, R. Soundararajan, and A. Bovik, "Making a "completely blind" image quality analyzer," *Signal Processing Letters, IEEE*, 20(3), 209–212, March 2013.

[31] M. A. Saad, A. C. Bovik, and C. Charrier, "Blind image quality assessment: A natural scene statistics approach in the dct domain," *IEEE Transactions on Image Processing*, 21(8), 3339–3352, 2012.

[32] E. Larson and D. Chandler, "Most apparent distortion: Full-reference image quality assessment and the role of strategy," *Journal of Electronic Imaging*, 19, 011006, 01 2010.

[33] A. Abdelhamed, S. Lin, and M. S. Brown, "A high-quality denoising dataset for smartphone cameras," in *IEEE Conference on Computer Vision and Pattern Recognition (CVPR)*, June 2018.

[34] R. Rajni and A. Anutam, "Image denoising techniques - an overview," *International Journal of Computer Applications*, 86, 12, 2013.

[35] X. Chu, L. Chen, and W. Yu, "NAFSSR: Stereo image super-resolution using nafnet," 2022.

[36] S. W. Zamir, A. Arora, S. H. Khan, H. Munawar, F. S. Khan, M.-H. Yang, and L. Shao, "Learning enriched features for fast image restoration and enhancement," *IEEE Transactions on Pattern Analysis and Machine Intelligence*, 2022.

[37] J. L. Roux and E. Vincent, "Consistent Wiener filtering for audio source separation," *IEEE Signal Processing Letters*, 20(3), 217–220, 2013.

[38] T. A. Parse, T. Awasthi, D. Yadav, and P. Joshi, "Qaad: Quality aware adaptive denoising," in *2024 11th International Conference on Signal Processing and Integrated Networks (SPIN)*, 2024, pp. 180–186.

[39] S. Chang, B. Yu, and M. Vetterli, "Adaptive wavelet thresholding for image denoising and compression," *IEEE Transactions on Image Processing*, 9(9), 1532–1546, 2000.

# Chapter 10

# Applications of Image Quality Assessment

IQA involves both subjective assessments (human evaluations) and objective metrics (algorithmic evaluations). While subjective assessments provide insights into perceived quality, objective metrics offer a faster and more cost-effective way to evaluate large datasets [1]. IQA has a wide range of applications across various fields. The following discussion provides some key areas where image quality assessment plays a vital role.

## 10.1   Image Compression

Image quality assessment plays a crucial role in image compression by ensuring that the compression algorithms maintain a balance between reducing file size and preserving visual quality. It helps in evaluating the quality of compressed images to ensure that the compression algorithms do not significantly degrade the visual quality [2]. Following are some key applications of image quality assessment in the domain of image compression:

1. **Evaluating Compression Algorithms:** Image quality assessment is crucial when evaluating compression algorithms, as it helps determine how well an algorithm preserves the visual quality of images while reducing their size. It is used to assess the performance of different compression algorithms, such as JPEG, PNG and HEVC. By comparing the quality of compressed images, researchers can determine which

algorithm provides the best trade-off between compression ratio and image quality [1]. The objective metrics such as Peak Signal-to-Noise Ratio (PSNR), Mean Squared Error (MSE) and Structural Similarity Index (SSIM) and subjective metric such as Mean Opinion Score (MOS) are some of the important metrics used to evaluate the performance of compression algorithms.

2. **Rate-Distortion Optimization (RDO):** In the process of compressing images, there is often a trade-off between the bit rate (the amount of data used to represent the image) and the distortion (loss of quality due to compression). RDO is a fundamental concept in image compression that aims to balance the trade-off between the bit rate and the distortion. It is a critical process in image compression and involves finding the right balance between the bit rate and the distortion. Achieving this balance ensures optimal visual quality while minimizing the data size. In video compression standards such as H.264 and HEVC, RDO is used extensively to optimize the encoding process. The encoder evaluates different encoding options and selects the one that provides the best trade-off between bit rate and distortion. Further, formats like JPEG and WebP use RDO principles to achieve efficient compression. Advanced codecs like JPEG 2000 and AVIF further enhance this by incorporating sophisticated IQA metrics. In real-time applications such as video streaming, RDO ensures that the visual quality is maintained while minimizing the bandwidth usage. This is crucial for providing a smooth viewing experience over varying network conditions.

So, IQA helps in optimizing this trade-off to achieve the best possible visual quality at a given bit rate [3]. It plays a crucial role in RDO by providing metrics to evaluate the visual quality of compressed images. The objective metrics such as Peak Signal-to-Noise Ratio (PSNR) and Structural Similarity Index (SSIM) and perceptual metrics such as Human Visual System (HVS) models are some of the important metrics of IQA which are used in getting the optimum values of RDO. Further, IQA metrics can be used to adaptively adjust the compression level based on the content of the image. For example, areas with high detail might be compressed less to preserve quality, while smoother areas can be compressed more. Hence, leveraging IQA in RDO, compression algorithms can achieve a balance that meets the specific needs of different applications, whether it's reducing storage requirements, optimizing for bandwidth or maintaining high visual fidelity.

3. **Artifact Detection:** Compression artifacts such as blocking (the presence of visible squares in the image, often seen in JPEG compression), blurring (the loss of sharpness and detail) and ringing (Halo effects around edges) are common while performing image compression. Image Quality Assessment (IQA) in artifact detection is essential for ensuring that images are free from distortions that can affect their usability and visual appeal. Compression can introduce various artifacts such as blocking, blurring and color shifts. IQA methods are used to detect and quantify these artifacts, helping developers to refine compression techniques to minimize these issues [1]. The effectiveness of artifact detection can vary based on the content of the image. High-detail images might require different approaches compared to simpler images.

   Objective IQA metrics, such as PSNR, SSIM and Mean Squared Error (MSE), are widely used in defect detection. Deep learning-based methods such as Convolutional Neural Networks (CNNs) have also been used recently for detecting and classifying artifacts in images. For example, an improved cascade region-based CNN can detect and localize multiple types of artifacts in endoscopic images [4]. Further, autoencoders (the neural networks that learn to encode and decode images) identify artifacts by comparing the input and output images. The perceptual metrics, such as Human Visual System (HVS) models which simulate human perception to evaluate how noticeable the artifacts are, also help in assessing the visual impact of artifacts more accurately.

4. **Standardization:** IQA is essential for developing and standardizing new compression methods. It plays a pivotal role in the development and standardization of new image compression methods. By providing a consistent way to measure image quality, IQA ensures that new compression techniques meet industry standards and user expectations [1]. For example, by providing a consistent and objective way to measure image quality, IQA ensures that new compression techniques not only meet industry standards but also align with user expectations. This is crucial for the widespread adoption and success of new compression methods, ultimately leading to better and more efficient ways to store and transmit images.

   IQA is essential in this context of Standardization due to several reasons. IQA provides standardized metrics that allow developers to benchmark the performance of new compression algorithms against existing ones. This ensures that any improvement in compression efficiency does not come at the cost of significant quality loss. Objective evaluation metrics such as PSNR, SSIM and VIF (Visual Information Fidelity) offer objective ways to measure image quality and these metrics provide consistent and repeatable results, making it easier to compare different compression methods. The industry standards for image compression, such as JPEG, HEVC and AV1, often include specific quality requirements. IQA helps ensure that new compression techniques comply with these standards by providing measurable quality benchmarks. Further, for a new compression method to be adopted widely, it often needs to be certified by industry bodies. IQA metrics are used in the certification process to verify that the new method meets the required quality standards.

Users expect compressed images to maintain a high level of visual quality. IQA metrics, especially those aligned with human perception like SSIM and MS-SSIM (Multi-Scale SSIM), help developers optimize compression algorithms to meet these expectations. By using IQA metrics during the development process, developers can receive immediate feedback on how changes to the compression algorithm affect image quality. This iterative process helps in fine-tuning the algorithm to balance compression efficiency and visual quality.

## 10.2   Medical Imaging

In medical diagnostics, high-quality images are crucial for accurate diagnosis and effective treatment. Image quality assessment is used to assess and improve the quality of medical images, such as MRI and CT scans, ensuring accurate diagnosis [2]. It ensures high-quality images for accurate diagnostics. Let us delve into the critical role of image quality in diagnostic accuracy:

1. **Enhanced Detection of Subtle Abnormalities:** High-quality medical images provide exceptional clarity and detail. This enables radiologists and clinicians to identify even the most subtle abnormalities within tissues, organs or structures. Whether it's a tiny tumor, a hairline fracture or early signs of disease, superior image quality ensures that nothing goes unnoticed.
2. **Mitigating Misdiagnoses:** Accurate diagnoses are essential for effective treatment planning. When image quality is compromised,

misinterpretations can occur. Misdiagnoses lead to unnecessary interventions, incorrect medications and delays in appropriate care. By prioritizing image quality, healthcare providers minimize the risk of such errors.

3. **Impact on Patient Outcomes:** The ultimate goal of medical imaging is to improve patient outcomes. High-quality images directly contribute to achieving this goal. When diagnoses are precise and timely, patients receive targeted treatments, experience fewer complications and have better prognoses. Whether it's a routine check-up or a complex medical condition, image quality significantly influences the patient's journey toward recovery.

Factors that affect the quality of a medical image include contrast, resolution, noise, artifacts and distortion [5]. Contrast differentiates between various tissues, while resolution (sharpness) determines the clarity of small details. Noise refers to unwanted random variations that can obscure details, and artifacts are unwanted alterations that can mislead diagnosis. Distortion involves the warping of the image, which can affect accuracy. Further, the quality assessment methods in medical imaging include both subjective and objective approaches. Subjective assessment relies on the radiographer's experience and training to evaluate image quality. On the other hand, objective metrics use tools like the Structural Similarity Index (SSIM) and Peak Signal-to-Noise Ratio (PSNR), which are specifically adapted for medical images, to provide quantifiable measures of image quality.

Quality Assurance (QA) in medical imaging is essential for maintaining consistent image quality and ensuring accurate diagnoses. Regular QA practices involve systematic equipment maintenance to prevent malfunctions and ensure optimal performance [6]. This includes routine checks, calibrations and timely repairs. Additionally, continuous training for radiographers and technicians is crucial, as it keeps them updated on the latest techniques and best practices. By combining these efforts, QA helps minimize errors, reduce variability in image quality and ultimately enhance patient care.

Advanced techniques are continually being developed to improve the quality of medical imaging. Portable fundus cameras, for instance, are increasingly used in telemedicine for eye examinations. These devices allow for high-quality imaging of the retina, facilitating remote diagnosis and monitoring of eye conditions. Susceptibility Weighted Imaging (SWI) is another advanced technique that enhances contrast in MRI scans, making it particularly useful for detecting small hemorrhages, vascular malformations and other abnormalities. Additionally, low-dose CT scans are designed to balance image quality with reduced radiation exposure, making them safer for patients while still providing the detailed images necessary for accurate diagnosis. These innovations collectively contribute to more effective and safer medical imaging practices. Improving image quality in medical imaging not only enhances diagnostic capabilities but also contributes to better patient outcomes.

## 10.3    Image Enhancement

Image enhancement techniques can significantly improve the visual quality of images compared to their original versions. Following are some of the important widely used image enhancement techniques.

1. **Deblurring:** This technique is essential for improving the sharpness of images that have been blurred due to camera shake, motion or out-of-focus capture. By applying algorithms that reverse the blurring effect, details become clearer, which is crucial in fields like forensic analysis and medical diagnostics.
2. **Contrast Adjustment:** Enhancing contrast helps in distinguishing different elements within an image. In medical imaging, for instance, adjusting contrast can make it easier to identify abnormalities in tissues. In photography, it can make images more visually appealing by highlighting the differences between light and dark areas.
3. **Noise Reduction:** Noise, which appears as random variations in brightness or color, can obscure important details. Techniques like median filtering, wavelet transforms and AI-based denoising are used to reduce noise, making the image cleaner and more informative. This is particularly important in low-light photography and in medical imaging modalities, such as MRI and CT scans.
4. **Resolution Enhancement:** Increasing the resolution of an image allows for finer details to be seen. Techniques such as super-resolution, which uses machine learning to upscale images, are used to enhance the resolution. This is beneficial in satellite imagery, medical imaging and any application where detail is critical.

5. **Color Enhancement:** Adjusting and enhancing colors can make images more realistic and visually pleasing. This involves correcting color imbalances and enhancing saturation and vibrancy. In digital media, this can make content more engaging, while in medical imaging, it can help in better visualization of tissues and organs.

6. **Image Brightening:** Enhancing the brightness of an image can reveal details that are hidden in shadows. This is particularly useful in security and surveillance, where it is important to see details in poorly lit conditions. In medical imaging, brightening can help in better visualization of structures within the body.

7. **Edge Enhancement:** This technique sharpens the edges within an image, making boundaries between different regions more distinct. This is useful in applications like object recognition and medical imaging, where clear delineation of structures is necessary

8. **Exposure Correction:** Image exposure correction in photography aims to regulate the inaccurate exposure settings of images by manipulating the underexposed and overexposed regions. This post-processing technique is particularly useful when the data for those regions are limited in the raw image [7]. A recent work in Ref. [8] introduces a novel method called decoupling and aggregating for image exposure correction. Another approach in Ref. [9] involves end-to-end exposure correction models that handle both underexposure and overexposure errors with a single model. Researchers have also formulated exposure correction as two main sub-problems, namely color enhancement and detail enhancement [10]. They propose a coarse-to-fine deep neural network (DNN) model that separately addresses each sub-problem.

9. **White Balance Adjustment:** White balance in photography refers to the process of correcting color casts in the photos. When an image is captured, the light source can introduce subtle color variations. For example, sunlight can range from blue (on cloudy days) to yellow (during the afternoon), while artificial lights like neon may add green or purple tints. Adjusting the white balance helps in achieving natural-looking colors.

These enhancements may sometimes inadvertently introduce noise and artifacts, etc., resulting in over-enhancement. To strike the right balance, it is crucial to optimize model structures and parameters, ensuring that enhancements maintain quality without compromising naturalness. IQA is used to evaluate the visual quality of an image to check the effectiveness of image enhancement techniques [11]. The application of image quality in image enhancement is pivotal across various fields, including medical imaging, photography, security and digital media.

## 10.4   Multimedia Applications

Image Quality Assessment (IQA) plays a crucial role in various multimedia applications, serving as an essential tool for evaluating the quality of images and videos across different platforms, such as streaming services, social media and broadcasting [12]. By ensuring that visual content meets high standards, IQA enhances user experience by delivering clear, detailed and visually appealing images. This is particularly important in streaming services where real-time quality adjustments can significantly improve viewing experiences and in social media where high-quality visuals are crucial for engagement. Additionally, in broadcasting, maintaining image quality is vital for delivering professional and reliable content. IQA's applications extend to various fields, including video compression, surveillance systems, photography, cinematography, augmented reality (AR), virtual reality (VR) and gaming, where it ensures that the visual quality meets the required standards, thereby improving overall user satisfaction and engagement. Some of the key applications are elaborated in the following:

1. **Video Streaming and Broadcasting:** In video streaming and broadcasting, Image Quality Assessment (IQA) is pivotal for ensuring that viewers receive the best possible visual experience. It helps maintain the quality of video streams by monitoring and adjusting the quality in real time [13]. IQA techniques are employed to monitor and evaluate the quality of video content in real time, allowing for adjustments that maintain high standards of clarity and detail. This is particularly important in live broadcasts, where any drop in quality can be immediately noticeable to viewers. By continuously assessing factors such as resolution, color accuracy and compression artifacts, IQA helps in delivering smooth and visually appealing streams. Additionally, in broadcasting, IQA ensures that the content meets professional standards, which is crucial for maintaining the credibility and reliability of the broadcast. Overall, IQA enhances user satisfaction by providing a consistent and high-quality viewing experience, whether it's for live sports, news or entertainment.

2. **Audio and Video Compression:** Image Quality Assessment (IQA) is integral to audio and video compression, ensuring that the compression process does not significantly degrade the perceptual quality of the content. It is used to evaluate the effectiveness of compression algorithms and ensures that the visual quality of multimedia content remains high despite compression [14]. In video compression, IQA techniques evaluate the effectiveness of algorithms by assessing factors such as resolution, color accuracy and the presence of compression artifacts. This helps in optimizing the balance between reducing file size and maintaining visual quality, which is crucial for efficient storage and transmission of video data [15, 16]. Similarly, in audio compression, IQA methods assess the clarity and fidelity of the audio after compression, ensuring that the sound quality remains high despite the reduced file size. By providing objective measures of quality, IQA enables the development of more efficient compression algorithms that deliver high-quality multimedia content while minimizing bandwidth and storage requirements [15].

3. **Surveillance Systems:** IQA is crucial in surveillance systems to ensure that the captured images and videos are clear and detailed, which is essential for effective monitoring and security. Surveillance images often suffer from various types of quality degradation due to factors like poor lighting, weather conditions and compression artifacts. IQA techniques are employed to continuously monitor and evaluate the quality of these images, identifying issues such as noise, blur and distortions [17]. By assessing these factors, IQA helps in maintaining high standards of image quality, which is vital for accurate recognition, detection and tracking in surveillance systems.

   Moreover, IQA can guide the optimization of camera settings and the development of more robust image processing algorithms, ensuring that surveillance footage remains reliable even under challenging conditions. This is particularly important in critical security applications, such as in airports, public spaces and high-security facilities, where the clarity and accuracy of surveillance images can significantly impact safety and security [18, 19]. Overall, IQA enhances the effectiveness of surveillance systems by providing consistent and high-quality visual data for monitoring and analysis.

   In summary, high resolution is crucial for identifying details, such as faces, license plates and other critical elements. Clarity ensures that the images are sharp and free from blurring. Further, proper lighting and contrast are essential for capturing clear images in various lighting conditions, including low light and high dynamic range scenarios. Surveillance footage often suffers from noise, especially in low-light conditions. Effective noise reduction techniques are necessary to maintain image quality. Also, surveillance systems often use compression to save storage space. However, excessive compression can introduce artifacts that degrade image quality. Balancing compression and quality is a key challenge.

4. **Photography and Cinematography:** Image quality assessment (IQA) in photography and cinematography is essential for ensuring high standards of visual content. It involves evaluating key aspects, such as resolution, sharpness, color accuracy, dynamic range and noise reduction. Both subjective methods, such as Mean Opinion Score (MOS), and objective methods, such as Full-Reference (FR) and No-Reference (NR) IQA, are used to assess quality. Recent advancements in this domain include the use of deep learning and AI to improve NR IQA and aesthetic quality models that evaluate compositional elements. These methods help photographers and filmmakers enhance post-processing, optimize camera and lens performance, and maintain high-quality streaming and broadcasting standards. They use IQA to ensure that their images and videos meet high-quality standards. Applications of image quality assessment in photography and cinematography include following:

   (a) **Post-Processing and Editing:** IQA helps photographers and filmmakers in post-processing by identifying areas that need enhancement, such as sharpness, color correction and noise reduction.

   (b) **Camera and Lens Testing:** IQA is used to evaluate the performance of cameras and lenses, ensuring they meet the required standards for professional use.

   (c) **Streaming and Broadcasting:** For cinematography, maintaining high image quality during streaming and broadcasting is crucial. IQA methods help in optimizing compression settings to balance quality and bandwidth.

   (d) **Aesthetic Quality Assessment:** In photography, aesthetic quality assessment involves evaluating compositional elements such as the rule of thirds, color harmony and depth of field. Computational models are increasingly used to automate this process.

5. **Augmented Reality (AR) and Virtual Reality (VR):** Image quality assessment (IQA) plays a crucial role in enhancing the user experience

in Augmented Reality (AR) and Virtual Reality (VR). In gaming and interactive experiences, high-quality AR and VR content enhances immersion and user engagement. IQA ensures that the visual elements are realistic and free from distracting artifacts. Further, AR and VR are used for educational and training purposes, where high-quality visuals are crucial for effective learning. IQA helps in maintaining the clarity and accuracy of instructional content. In fields such as architecture, engineering and medical training, AR and VR simulations require high visual fidelity to accurately represent real-world scenarios. IQA ensures that these simulations are reliable and effective.

Particularly in AR, IQA ensures that virtual elements seamlessly blend with the real-world environment, maintaining high visual fidelity and minimizing artifacts that could disrupt the immersive experience [20]. Techniques such as Confusing Image Quality Assessment (CFIQA) evaluate the perceptual quality of superimposed images, helping achieve better Quality of Experience (QoE) by considering the interaction between virtual and real scenes [21]. The following discussion provides more details of the application of image quality assessment (IQA) in Augmented Reality (AR):

(a) **Seamless Integration:** In AR, virtual objects are superimposed onto the real world. IQA ensures that these virtual elements blend seamlessly with the real environment. This involves assessing the alignment, color consistency and lighting of virtual objects to match the real-world scene.

(b) **Perceptual Quality:** Techniques like Confusing Image Quality Assessment (CFIQA) evaluate how well the virtual elements integrate with the real world from a perceptual standpoint. This includes minimizing artifacts such as ghosting, blurring or misalignment that can disrupt the immersive experience.

(c) **Real-Time Processing:** AR applications often require real-time processing of images and videos. IQA methods help in optimizing the performance of AR systems by ensuring that the quality of the augmented content is maintained without causing significant delays or computational overhead.

In VR, IQA is essential for assessing the quality of omnidirectional images and videos, which are critical for creating realistic and immersive environments. Advanced methods, such as VR IQA with adversarial learning, predict quality scores based on human perception, ensuring that the virtual content is both visually appealing and comfortable for users [22]. Automated visual quality assessment using machine learning also helps in optimizing the production of 3D assets, reducing costs and improving the overall quality of VR experiences [23]. By implementing robust IQA methods, AR and VR systems can deliver high-quality, immersive experiences that meet the expectations of users across various applications, from entertainment and education to training and industrial simulations. The following discussion provides more details of the application of image quality assessment (IQA) in Virtual Reality (VR):

(a) **Omnidirectional Content:** VR involves 360-degree images and videos that create an immersive environment. IQA methods assess the quality of these omnidirectional contents to ensure that they are free from stitching errors, blurring and other artifacts that can break the immersion.

(b) **Human Perception:** Advanced IQA methods, such as VR IQA with adversarial learning, predict quality scores based on human perception. These methods consider factors like visual comfort, depth perception and motion artifacts to ensure that the VR experience is both visually appealing and comfortable for users.

(c) **3D Asset Optimization:** In VR, high-quality 3D models and textures are crucial for creating realistic environments. Automated visual quality assessment using machine learning helps in optimizing the production of these assets, ensuring that they meet the required quality standards while reducing production costs.

6. **Gaming:** In game development, image quality assessment (IQA) is a critical process used to ensure that the graphics are of high quality, thereby enhancing the overall gaming experience. High-quality graphics are essential for creating immersive and visually appealing game environments that captivate players. IQA methods are employed to evaluate various aspects of game graphics, such as resolution, texture detail, lighting and color accuracy. By assessing these elements, developers can identify and rectify any visual imperfections that might detract from the gaming experience.

For instance, IQA helps in optimizing textures to ensure they are sharp and detailed, which is crucial for creating realistic environments. It also evaluates lighting effects to ensure they are consistent and enhance the mood and atmosphere of the game. Additionally, IQA methods are used to detect and reduce visual artifacts, such as aliasing and compression artifacts, which can degrade the visual quality of the game.

Advanced IQA techniques, including those based on machine learning and deep learning, are increasingly used to automate the assessment process. These techniques can analyze large volumes of graphical data and provide insights into areas that need improvement. By implementing robust IQA methods, game developers can ensure that their games not only look stunning but also provide a smooth and enjoyable experience for players, ultimately leading to higher player satisfaction and engagement. Key aspects of IQA in game development are elaborated in the following:

(a) **Resolution and Texture Quality:** High resolution and detailed textures are fundamental for creating realistic and immersive game environments. IQA methods evaluate the sharpness and clarity of textures, ensuring that they are free from blurring and pixelation. This is particularly important for high-definition displays and VR gaming, where players can notice even minor imperfections.

(b) **Lighting and Shadows:** Proper lighting and shadow effects are crucial for setting the mood and atmosphere in a game. IQA assesses the consistency and realism of lighting, ensuring that shadows are accurately cast and that lighting transitions are smooth. This helps in creating a more believable and engaging environment.

(c) **Color Accuracy and Consistency:** Accurate color reproduction is essential for maintaining the visual integrity of a game. IQA methods evaluate the color balance and consistency across different scenes and levels, ensuring that colors are vibrant and true to the intended design.

(d) **Frame Rate and Motion Smoothness:** A high and stable frame rate is vital for a smooth gaming experience. IQA assesses the frame rate performance, identifying any drops or stutters that could affect gameplay. This is especially important in fast-paced games where smooth motion is critical for player performance.

(e) **Visual Artifacts:** IQA helps in detecting and reducing visual artifacts, such as aliasing, compression artifacts and screen tearing. These artifacts can significantly degrade the visual quality and immersion of a game. Techniques such as antialiasing and advanced compression algorithms are used to minimize these issues.

Advanced IQA methods leverage machine learning and AI to automate the assessment process. These techniques can analyze large datasets of game graphics, learning to identify patterns and features that correlate with high-quality visuals. This allows for more efficient and accurate quality assessments. In modern game development, real-time IQA is used to monitor and adjust graphics quality dynamically. This ensures that the game maintains high visual standards even under varying hardware conditions and performance constraints. Some IQA methods focus on the player's perception of quality. By incorporating user feedback and subjective evaluations, developers can fine-tune the visual elements to better meet player expectations and preferences.

By implementing robust IQA methods, game developers can create visually stunning games that provide an immersive and enjoyable experience for players. This not only enhances player satisfaction but also contributes to the commercial success of the game. Some of the practical applications of image quality assessment in gaming are elaborated in the following:

(a) **Game Testing and Optimization:** During the development phase, IQA is used extensively in game testing to identify and fix graphical issues. This includes optimizing textures, lighting and effects to ensure they perform well across different hardware configurations.

(b) **Post-Processing Effects:** IQA helps in evaluating and enhancing post-processing effects, such as bloom, depth of field and motion blur. These effects add to the visual richness of a game but need to be carefully balanced to avoid overloading the visual experience.

(c) **Cross-Platform Consistency:** With games being released on multiple platforms, IQA ensures that the visual quality is consistent across different devices, from high-end gaming PCs to consoles and mobile devices. This involves adjusting graphical settings and assets to suit the capabilities of each platform.

## 10.5   Image Restoration

Image Restoration (IR) aims to recover high-quality images from degraded versions, which can be affected by factors, such as noise, blur or compression artifacts. Image Quality Assessment (IQA) methods are essential for evaluating how effectively these degradations have been mitigated by IR algorithms. They provide quantitative metrics that facilitate the comparison of different IR techniques and guide their development.

Therefore, IQA methods are crucial for assessing the effectiveness of IR algorithms. They play a vital role in the development and evaluation of IR algorithms by quantifying the restoration quality of images. This quantification is essential for improving and comparing various IR techniques [24]. IQA methods help evaluate the performance of IR techniques, which aim to recover high-quality images from degraded versions [25].

Generative Adversarial Networks (GANs) have revolutionized IR by producing visually appealing results. Recent IR methods, especially those based on generative adversarial networks, have significantly improved visual performance. However, they also pose challenges for quantitative evaluation. Traditional IQA methods often struggle to fairly evaluate these advanced algorithms [24, 26] and traditional IQA metrics like Peak Signal-to-Noise Ratio (PSNR) and Structural Similarity Index (SSIM) often fail to capture the perceptual quality of images restored by GANs. To address the limitations of traditional IQA methods, the PIPAL dataset was created. This dataset includes images processed by various IR algorithms, including GAN-based methods, and provides subjective scores based on human judgments. The dataset contains over 1.13 million human ratings, making it one of the largest and most comprehensive resources for evaluating IQA methods. Researchers use this dataset to train and benchmark new IQA models that better align with human perception.

Metrics like PSNR, SSIM and Learned Perceptual Image Patch Similarity (LPIPS) provide automated ways to evaluate image quality. They are computationally efficient and can be used for large-scale evaluations. However, they may not always correlate well with human perception, especially for complex distortions introduced by advanced IR methods. So, while objective metrics are useful, subjective assessment methods, such as Mean Opinion Scores (MOS) and Differential Mean Opinion Scores (DMOS), are often employed to evaluate the perceptual quality of restored images. These methods rely on human judgments to provide a more accurate assessment of image quality [27]. Researchers are continuously working on enhancing IQA methods to better align with the advancements in IR algorithms. Some of the recent advancements in this domain include the following:

1. **Deep Learning-based IQA:** Models such as the Space Warping Difference Network (SWDN) leverage deep learning to better capture perceptual differences in images. These models are trained on large datasets

such as PIPAL to improve their accuracy in assessing GAN-based distortions.

2. **Perceptual Metrics:** Metrics such as LPIPS and Fréchet Inception Distance (FID) are designed to evaluate the perceptual similarity between images. They use deep neural networks to capture high-level features that are more aligned with human perception.

## 10.6   Printing

Image Quality Assessment (IQA) for printing is essential to ensure that the final printed output meets the desired standards. The key quality attributes which are considered during the evaluation include sharpness, color accuracy, lightness and contrast and artifacts. Here, sharpness measures the clarity and detail of the printed image. It is crucial for text readability and fine details in images. Further, color accuracy ensures that the colors in the printed image match the original digital image. This is vital for applications such as advertising and photography where color fidelity is important. Lightness and contrast evaluate the brightness and contrast levels to ensure that the printed image is neither too dark nor too light. Artifacts check for any unwanted elements like banding, ghosting or color bleeding that can degrade the quality of the printed image.

Quality assessment is done using both subjective and objective metrics. In subjective metrics, human observers are often used to evaluate print quality through subjective methods. This involves rating the quality of printed images based on visual inspection. Subjective assessment is considered the gold standard as it directly reflects human perception. Objective metrics used for evaluation includes the following:

1. **DPI (Dots Per Inch):** Measures the resolution of the printed image. Higher DPI values generally indicate better quality as they provide more detail.
2. **SSIM (Structural Similarity Index):** Compares the structural information between the original and printed images to assess quality.
3. **Color Metrics:** Metrics such as Delta E measure the difference between the colors in the original and printed images to ensure color accuracy.

Print quality analysis systems are also employed in quality control processes to ensure that the printed output meets the required standards. These systems break down images into dots, lines and texts to perform specialized measurements. They check for issues like missing lines, uniformity of patterns and color bleeding.

# 10.7   Scanning

Image Quality Assessment (IQA) in scanning is crucial for ensuring that the scanned images meet the desired quality standards. Key quality attributes involved in the assessment include the following:

1. **Resolution:** The resolution of a scanner, typically measured in DPI (dots per inch), determines the level of detail captured in the scanned image. Higher resolution settings capture more detail but result in larger file sizes.
2. **Sharpness:** It measures the clarity and detail of the scanned image. Sharpness is essential for text readability and fine details in images [28].
3. **Color Accuracy:** It ensures that the colors in the scanned image match the original document. This is vital for applications like archiving and digital reproduction of artworks [28].
4. **Noise:** It refers to random variations in brightness or color information in the scanned image. Lower noise levels indicate higher quality [29].
5. **Artifacts:** It checks for any unwanted elements like banding, ghosting or color bleeding that can degrade the quality of the scanned image [28].

Here also, quality assessment is done using both subjective and objective metrics. In subjective metrics, human observers are often used to evaluate scan quality through subjective methods. This involves rating the quality of scanned images based on visual inspection. Subjective assessment is considered the gold standard as it directly reflects human perception [28]. The objective metrics used for evaluation in case of assessing the quality of a scanned page include the following:

1. **Noise Level:** It measures the amount of noise in the scanned image. Lower noise levels are preferred for higher quality [29].
2. **Contrast-to-Noise Ratio (CNR):** It evaluates the contrast of the image relative to the noise. Higher CNR values indicate better quality [28].
3. **Modulation Transfer Function (MTF):** It assesses the scanner's ability to reproduce fine details. Higher MTF values indicate better sharpness [28].

4. **Non-Uniformity Index (NUI):** It measures the uniformity of the scanned image. Lower NUI values indicate more uniform images.

# References

[1]    S. Jamil, "Review of image quality assessment methods for compressed images," *Journal of Imaging*, 10, 113, 2024.

[2]    M. Khosravy, N. Patel, N. Gupta, and I. K. Sethi, "Image quality assessment: A review to full reference indexes," in *Recent Trends in Communication, Computing, and Electronics*, 2018, pp. 279–288.

[3]    X. Zhang, W. Lin, S. Wang, J. Liu, S. Ma, and W. Gao, "Fine-grained quality assessment for compressed images," *IEEE Transactions on Image Processing*, 28(11), pp. 5416–5427, 2019.

[4]    L. Zhang, L. Zhang, X. Mou, and D. Zhang, "Fine-tuning convolutional neural networks for visual quality assessment," *IEEE Transactions on Image Processing*, 28(3), 1126–1137, 2019.

[5]    M. Dumas, M. L. Rosa, J. Mendling, and H. A. Reijers, *Fundamentals of Business Process Management*. Springer, 2018. Available: https://link.springer.com/book/10.1007/978-3-662-56509-4

[6]    G. James, D. Witten, T. Hastie, and R. Tibshirani, *An Introduction to Statistical Learning: with Applications in R*. Springer, 2021. Available: https://link.springer.com/book/10.1007/978-1-0716-1418-1

[7]    M. Parab, A. Bhanushali, P. Ingle, and B. N. P. Kumar, "Image enhancement and exposure correction using convolutional neural network," *SN Computer Science*, 4(204), 1–14, 2023.

[8]    Y. Wang, L. Peng, L. Li, Y. Cao, and Z.-J. Zha, "Decoupling-and-aggregating for image exposure correction," in *Proceedings of the IEEE/CVF Conference on Computer Vision and Pattern Recognition (CVPR)*. IEEE, 2023, pp. 18 115–18 124.

[9]    F. I. Eyiokur, D. Yaman, H. K. Ekenel, and A. Waibel, "Exposure correction model to enhance image quality," in *Proceedings of the IEEE/CVF Conference on Computer Vision and Pattern Recognition*, 2022, pp. 676–686.

[10]   M. Afifi, K. G. Derpanis, B. Ommer, and M. S. Brown, "Learning multi-scale photo exposure correction," in *Proceedings of the IEEE/CVF Conference on Computer Vision and Pattern Recognition (CVPR)*. IEEE, 2021, pp. 676–686.

[11]   K. Ohashi, Y. Nagatani, M. Yoshigoe, K. Iwai, K. Tsuchiya, A. Hino, Y. Kida, A. Yamazaki, and T. Ishida, "Applicability evaluation of full-reference image quality assessment methods for computed tomography

images," *Journal of Imaging Informatics in Medicine*, 36, 2623–2634, 2023.

[12] K. Ma and Y. Fang, "Image quality assessment in the modern age," *CoRR*, vol. abs/2110.09699, 2021. Available: https://arxiv.org/abs/2110.09699

[13] K. Ma and Y. Fang, "Image quality assessment in the modern age," in *Proceedings of the 29th ACM International Conference on Multimedia (MM 2021)*, 2021, pp. 5664–5666.

[14] X. Yang, T. Wang, and G. Ji, "Image quality assessment via multiple features," *Multimedia Tools and Applications*, 81, 5459–5483, 2022. Available: https://link.springer.com/article/10.1007/s11042-021-11788-x

[15] S. Wen and J. Wang, "A strong baseline for image and video quality assessment," *arXiv preprint arXiv:2111.07104*, 2021. Available: https://arxiv.org/abs/2111.07104

[16] K. Seshadrinathan and A. C. Bovik, "Image and video quality assessment," in *Encyclopedia of Multimedia*. Springer, 2008. Available: https://link.springer.com/referenceworkentry/10.1007/978-0-387-78414-4_341

[17] Z. Ye, X. Ye, and Z. Zhao, "Hybrid no-reference quality assessment for surveillance images," *Information*, 13(12), 588, 2022. Available: https://www.mdpi.com/2078-2489/13/12/588

[18] Z. Ye, X. Ye, and Z. Zhao, "Hybrid no-reference quality assessment for surveillance images," *Information*, 13(12), 588, 2022. Available: https://doi.org/10.3390/info13120588

[19] Y. Zhang, X. Wang, H. Li, and J. Liu, "Image quality assessment for video surveillance system," in *Advances in Multimedia Information Processing PCM 2017*. Cham: Springer International Publishing, 2018, pp. 838–846. Available: https://link.springer.com/chapter/10.1007/978-3-319-77383-4_82

[20] H. taek Lim, H. G. Kim, and Y. M. Ro, "Vr iqa net: Deep virtual reality image quality assessment using adversarial learning," *CoRR*, vol. abs/1804.03943, 2018. Available: https://arxiv.org/abs/1804.03943

[21] H. Duan, X. Min, Y. Zhu, G. Zhai, X. Yang, and P. L. Callet, "Confusing image quality assessment: Toward better augmented reality experience," *IEEE Transactions on Image Processing*, 31, 7206–7221, 2022.

[22] B. Roullier, F. McQuade, A. Anjum, C. Bower, and L. Liu, "Automated visual quality assessment for virtual and augmented reality based digital twins," *Journal of Cloud Computing*, vol. 13, p. Article number: 51, 2024. Available: https://journalofcloudcomputing.springeropen.com/articles/10.1186/s13677-024-00616-w

[23] J. Hettig, S. Engelhardt, C. Hansen, and G. Mistelbauer, "Ar in vr: assessing surgical augmented reality visualizations in a steerable virtual reality environment," *International Journal of Computer Assisted*

*Radiology and Surgery*, 13, 1717–1725, 2018. Available: https://link.sprin ger.com/article/10.1007/s11548-018-1825-4

[24]  J. Gu, H. Cai, H. Chen, X. Ye, J. S. J. Ren, and C. Dong, "Image quality assessment for perceptual image restoration: A new dataset, benchmark and metric," *CoRR*, vol. abs/2011.15002, 2020. Available: https://arxiv.org/ abs/2011.15002

[25]  Z. Wang, "Objective image quality assessment: Facing the real-world challenges," in *Proc. IS&T Int'l. Symp. on Electronic Imaging: Image Quality and System Performance XIII*, 2016.

[26]  J. Gu, H. Cai, H. Chen, X. Ye, J. S. Ren, and C. Dong, "PIPAL: a large-scale image quality assessment dataset for perceptual image restoration," in *European Conference on Computer Vision (ECCV)*.    Springer, 2020, pp. 633–651.

[27]  Z. Wang, J. Zhuang, S. Ye, N. Xu, J. Xiao, and C. Peng, "Image restoration quality assessment based on regional differential information entropy," *Entropy*, vol. 25, no. 1, p. 144, 2023. Available: https://www.mdpi.com/10 99-4300/25/1/144

[28]  H.-J. Bae, J.-H. Kim, S.-H. Kim, S.-H. Lee, S.-H. Lee, and S.-H. Lee, "A comprehensive assessment of physical image quality of five different scanners for head ct imaging," *PLOS ONE*, 16(1), e0245374, 01 2021. Available:    https://journals.plos.org/plosone/article?id=10.1371/journal. pone.0245374

[29]  Y. Zhang, Y. Zhang, Y. Zhang, Y. Zhang, and Y. Zhang, "Fully automated image quality evaluation on patient ct," *PLOS ONE*, 17(7), e0271724, 07 2022. Available: https://journals.plos.org/plosone/article?id=10.1371/jour nal.pone.0271724

# Chapter 11

# Challenges in Image Quality Assessment and Breakthroughs

## 11.1  Challenges in Image Quality Assessment

While image quality assessment has made significant strides, there are many challenges where further research and innovation are required [1]. Some of the important challenges are discussed here.

### 11.1.1  *Subjectivity and variability in human perception*

One of the primary challenges in IQA is the subjective nature of human perception. Subjective assessments can vary widely among different observers and different individuals may perceive the quality of the same image differently, making it difficult to develop universally accepted metrics [1].

These challenges exist due to incomplete knowledge about human visual perception. The existing IQA techniques are designed or trained with insufficient subjective data, resulting in difficulty handling the complexity and diversity of real-world digital content. To address this, researchers conduct controlled perception experiments to collect reliable subjective data that faithfully reflect human behavioral responses to distortions in visual signals. A recent study conducted in Ref. [2] investigates how quality perception is affected by different categories of images and various types and levels of distortions. The researchers collected subjective ratings in a controlled lab environment and created a publicly available database for calibrating and validating IQA algorithms.

## 11.1.2   *Computational complexity*

Computational complexity plays a crucial role in designing effective IQA methods. Researchers continue to explore innovative approaches to balance accuracy, efficiency and scalability in assessing image quality.

As we know, objective IQA techniques attempt to predict quality of an image as perceived by human observers. These methods can be broadly categorized into two types based on the availability of a reference image. In Full-Reference IQA (FR-IQA), image quality is assessed by comparing the distorted image with a reference image and providing a quality score based on the distortion between the two images. In No-Reference IQA (NR-IQA), image quality is estimated without a reference image. These methods directly work on the distorted image to analyze it to predict its quality [3]. These objective quality assessment methods can be computationally intensive, particularly for real-time applications.

Among the two types of objective image quality methods, specially no-reference IQA methods face challenges due to the increasing complexity of networks and components used to improve performance. This complexity becomes particularly prominent when we deal with high-resolution images in different real-world applications.

It is seen that most of the high-resolution images exhibit high spatial redundancy and existing no-reference IQA methods lack in handling this complexity effectively. A solution in this direction is proposed in Ref. [4] where the proposed framework addresses the limitations of existing no-reference-IQA methods by leveraging compressive sampling and deep learning. It samples the image at an arbitrary ratio, to exploit the spatial redundancy, whereas vision transformer with Adaptive Embedding Module (AEM) extracts deep features from uniform-sized measurements. Subsequently, Dual Branch (DB) allocates weights for different image patches and predicts the final quality score.

## 11.1.3   *Diverse image content*

Different types of images, such as natural scenes and medical images, may require different assessment techniques. We know that the human perception of image quality varies based on context, culture and personal preferences. Capturing this diversity in subjective ratings is challenging. Similarly, extracting relevant features for an image quality assessment technique from diverse content of the images is non-trivial. For example, some features, such as texture and color, may be more critical for certain types of images. Further, diverse content introduces bias in image quality assessment models and ensuring fairness across content types is critical.

Balancing accuracy and computational complexity in image quality assessment methods is essential. Efficiently handling and assessing the quality of diverse images at a large scale can be achieved with the use of optimized algorithms. Further, it is seen that image quality assessment models generalize well when diverse training data is used. Hence, curating datasets with varied content, for example, with respect to natural scenes, portraits, abstract art, etc., can ensure the robustness of quality assessment.

### 11.1.4  *Lack of complete perceptual models*

The development of comprehensive perceptual models for natural images remains an ongoing challenge. These models play a crucial role in understanding how humans perceive image quality. The following discussion provides a discussion on factors imposing the challenge:

1. **Color Perception:** Color is a fundamental aspect of visual perception. However, creating accurate models that account for color perception across diverse scenes and lighting conditions is complex. Challenges include handling color constancy (the ability to perceive consistent colors despite varying illumination) and color adaptation (how our eyes adjust to different lighting environments).
2. **Texture and Patterns:** Texture perception involves analyzing repetitive patterns and variations in intensity or structure within an image. Capturing texture perception requires models that consider spatial frequency (high vs. low), orientation and scale. These factors impact our perception of fine details and overall texture.
3. **Spatial Frequency and Resolution:** Spatial frequency refers to the rate of change in intensity across an image. High spatial frequencies correspond to fine details, while low frequencies represent broader features. Models must account for how our visual system processes

different spatial frequencies. For instance, we're more sensitive to mid-range frequencies.

4. **Contrast Sensitivity:** Contrast sensitivity describes our ability to detect differences in luminance (brightness) between adjacent regions. Models need to incorporate contrast sensitivity functions, which vary with spatial frequency. Our perception is nonlinear, emphasizing certain contrasts more than others.

5. **Contextual Interactions:** Our perception is influenced by context. Surrounding elements impact how we perceive individual features. Models should consider contextual interactions, such as simultaneous contrast (where adjacent colors affect each other) and figure-ground segregation.

6. **Adaptation and Fatigue:** Our visual system adapts to prolonged exposure to specific stimuli. Adaptation affects our perception of subsequent images. Models should account for adaptation and fatigue effects, especially when assessing image quality over time.

To summarize, creating comprehensive perceptual models involves balancing these factors while considering the intricacies of human vision. Researchers continue to explore novel approaches, including deep learning, to improve our understanding of image quality perception.

### 11.1.5 *Compound and multiple distortions*

Images often suffer from multiple distortions. For example, an image may contain compression artifacts and at the same time may be noisy. Understanding their combined impact on perceived quality is complex and interactions between different distortions need extensive exploration. The existing image quality assessment technique struggles to handle compound distortions, that is, the presence of multiple types of distortions simultaneously. Developing models that can accurately assess quality under these conditions remains a challenge. Further, traditional image quality assessment models are mostly designed for common image distortions, such as compression and blur. However, real-world images exhibit diverse and unconventional distortions. Handling non-traditional distortions, such as artistic filters and stylization, poses a challenge.

## 11.2   Potential Breakthroughs and Innovations in Image Quality Assessment

The field of image quality assessment is rapidly evolving, with several potential breakthroughs and innovations on the horizon. These advancements promise to enhance the accuracy, efficiency and applicability of image quality assessment across various domains. Some key areas where significant progress is expected are elaborated in the following.

### 11.2.1   *Deep learning and artificial intelligence integration*

Deep learning and artificial intelligence (AI) are revolutionizing IQA by enabling the development of models that can learn from vast amounts of data. These models can automatically identify and quantify image distortions, leading to more accurate and reliable assessments. Innovations in this area include development in following fields:

**Convolutional Neural Networks (CNNs):** These networks are particularly effective in image processing tasks and are being used to develop advanced IQA models. CNNs have significantly advanced the field of image quality assessment by providing powerful tools for automatically learning and extracting features from images. Advanced IQA models using CNNs represent a significant leap forward in the field of image quality assessment. By leveraging the powerful feature extraction capabilities of CNNs, these models provide more accurate, robust and perceptually aligned assessments of image quality. Following are some of the notable advanced IQA models that use CNNs for image quality assessment:

1. **No-Reference Image Quality Assessment (NR-IQA) with CNNs:** No-Reference IQA models aim to assess image quality without requiring a reference image. These models are particularly challenging because they must infer quality based solely on the distorted image. CNNs have been effectively used in NR-IQA due to their ability to learn complex features from images. One such model is Multi-Scale Residual CNN with Attention Mechanism (MsRCANet) [5]. This model uses multi-scale feature extraction to capture detailed information from images at different resolutions. It incorporates a residual learning strategy and a channel attention mechanism to enhance feature extraction

and improve the accuracy of quality assessment. Another approach where CNNs are used for image quality assessment is the hybrid CNN-transformer model [6]. This model combines CNNs with transformers to leverage both local and non-local features. CNNs capture local structure information, while transformers handle non-local dependencies, providing a comprehensive assessment of image quality

2. **Full-Reference Image Quality Assessment (FR-IQA) with CNNs:** Full-Reference IQA models compare a distorted image to a high-quality reference image. CNNs enhance these models by learning to extract relevant features that highlight differences between the reference and distorted images. One such approach is deep feature-based FR-IQA. This model uses deep features extracted from pre-trained CNNs (such as VGG and ResNet) to compare the reference and distorted images. The deep features capture high-level perceptual information, leading to more accurate quality assessments.

3. **Hybrid Models:** These models combine traditional IQA metrics with deep learning features to provide a more comprehensive assessment of image quality. These models leverage the strengths of both approaches to improve accuracy and robustness. For example, in multi-stage hybrid models, first traditional metrics such as PSNR and SSIM is used for an initial assessment, which is followed by deep learning models to refine the quality score. This multi-stage approach enhances the robustness and accuracy of the assessment.

**Generative Adversarial Networks (GANs):** GANs can generate high-quality images and are being explored for their potential in enhancing IQA by creating reference images for comparison. Generative adversarial networks have revolutionized the field of image generation and have also found significant applications in image quality assessment. GANs consist of two neural networks, a generator and a discriminator, that are trained simultaneously through adversarial processes. This unique architecture allows GANs to generate high-quality images and assess the quality of generated images effectively. The applications of GANs in IQA are many out of which a few main are discussed in the following:

1. **Generated Image Quality Assessment:** It focuses on evaluating the quality of images generated by GANs. Traditional IQA methods may not be suitable for generated images due to unique artifacts introduced by GANs. Generated image quality assessment algorithms are

designed to assess the realism and quality of each generated image [7]. In learning-based generated image quality assessment, deep learning models are used to learn the characteristics of high-quality images and assess the quality of generated images based on these learned features. In data-based generated image quality assessment, statistical analysis of image data is carried out to evaluate the quality of generated images.

2. **Improving GAN Training:** IQA metrics can be used to improve the training of GANs. By incorporating IQA feedback, GANs can be trained to produce higher-quality images. For example, online hard negative mining can be used to focus the training on challenging examples, improving the overall quality of generated images [7].

3. **Enhancing Perceptual Quality:** GANs can be used to enhance the perceptual quality of images in various applications, such as image super-resolution, image denoising and image inpainting. By training GANs with perceptual loss functions that incorporate IQA metrics, the generated images can achieve higher visual quality.

### 11.2.2   *Real-T=time image quality assessment*

The demand for real-time image quality assessment is growing, especially in applications such as video streaming, gaming and surveillance. Efficient algorithms are essential for real-time image quality assessment, enabling quick and accurate evaluations across various applications. By leveraging advancements in deep learning, edge computing and parallel processing, these algorithms can meet the demands of real-time IQA, ensuring high-quality visuals in fields such as photography, medical imaging and autonomous driving.

Innovations in this area focus on reducing computational complexity and improving processing speeds. Different techniques that are developed in this area include development of new and efficient algorithms to perform IQA with lower computational requirements, making real-time assessment feasible. Efficient algorithms are essential for real-time image quality assessment due to the need for quick and accurate evaluations. SSIM is a popular metric for assessing image quality by comparing structural information. Optimized implementations of SSIM, such as using integral images or parallel processing, significantly speed up the calculation, making it feasible for real-time use. CNNs are widely used for image quality assessment due to their ability to learn and extract features from images [8]. Efficient CNN architectures, such as MobileNet and

EfficientNet, are designed to be lightweight and fast, making them suitable for real-time applications. Multi-scale and multi-level analysis-based image quality assessment involves evaluating images at different resolutions and levels of detail. This approach is improved by maintaining the accuracy of IQA using techniques such as Laplacian pyramids and wavelet transforms. Further, efficient data structures, such as integral images and summed-area tables, can speed up the computation of image quality metrics. These structures allow for quick aggregation of pixel values, reducing the computational complexity of IQA algorithms. Perceptual hashing algorithms generate a compact representation of an image that captures its perceptual quality. These hashes can be quickly compared to assess image quality, making them suitable for real-time applications. Decentralized processing helps in improving the computational speed. By performing IQA closer to the data source, edge computing reduces latency and bandwidth usage. This is particularly beneficial for applications such as autonomous driving, where quick and accurate image quality assessment is crucial for safety.

Hardware acceleration is crucial for real-time image quality assessment due to the computational intensity of the algorithms involved. In hardware acceleration, specialized hardware, such as Graphical Processing Units (GPUs) and Field Programmable Gate Arrays (FPGAs), are utilized to accelerate image quality assessment processes. Further, parallel and distributed power of computing resources has the potential to speed up the IQA algorithms. These hardware accelerators are capable of handling large-scale image data and complex computations in real time. For example, graphics processing units are highly effective for accelerating IQA algorithms due to their parallel processing capabilities. GPUs can handle multiple operations simultaneously, making them ideal for the intensive computations required in IQA. The GPU implementation of Most Apparent Distortion (MAD) Algorithm has shown significant speedups [9]. For instance, a single-GPU implementation can achieve a $24\times$ speedup, while a multi-GPU setup can reach up to $33\times$ speedup compared to a CPU implementation. Understanding the interaction between the program and the underlying hardware is essential for identifying and resolving performance bottlenecks. Techniques such as blocking, which map well to GPU memory hierarchies, can further enhance performance due to providing an optimized implementation [10]. General-Purpose GPU (GPGPU) involves the use of GPUs for general-purpose computations beyond graphics rendering. This approach leverages the massive parallelism of GPUs to accelerate various computational tasks, including IQA. Implementation of log-Gabor decomposition on GPUs has been shown significantly improving the statistical computations involved in IQA. Pre-computation and caching of filters further enhance performance [9].

Field-programmable gate arrays offer another avenue for accelerating IQA algorithms. FPGAs can be customized to perform specific tasks efficiently, making them suitable for real-time applications. Using OpenCL, IQA algorithms can be implemented on FPGAs to achieve significant performance improvements. For example, the most apparent distortion algorithm has been successfully accelerated on FPGAs, demonstrating the potential of this approach [11].

It is observed that identifying and addressing performance bottlenecks is crucial for optimizing IQA algorithms. Techniques such as using integral images for statistical computations, procedure expansion and strength reduction can lead to substantial speedups. Hence, understanding the micro-architectural details of the hardware can help in optimizing the implementation of IQA algorithms. This includes analyzing the runtime of different kernels and optimizing memory access patterns.

To summarize, we see that hardware acceleration plays a vital role in enhancing the performance of image quality assessment algorithms. By utilizing GPUs, FPGAs and GPGPU computing, significant speedups can be achieved, making real-time IQA feasible for various applications. As technology continues to evolve, further advancements in hardware acceleration will likely drive the development of even more efficient and accurate IQA methods.

### 11.2.3 *Multi-modal image quality assessment*

Combining information from multiple modalities, such as visual, auditory and textual data, can provide a more comprehensive assessment of image quality. Multi-modal image quality assessment (IQA) leverages multiple types of data, such as visual and textual information, to evaluate the quality of images more comprehensively. This approach enhances the accuracy and robustness of IQA models by integrating diverse sources of information. Innovations in multi-modal image quality assessment are being tried in multiple facets. Following are a few important directions:

**Cross-Modal Learning:** Cross-modal learning in multi-modal image quality assessment represents a significant advancement in the field of IQA.

By integrating visual and textual information, these models can provide more accurate, robust and comprehensive evaluations of image quality. It involves techniques that leverage information from different modalities to improve the accuracy of image quality assessment models. It involves integrating information from different modalities, such as visual and textual data, to enhance the accuracy and robustness of IQA models. This approach leverages the strengths of each modality to provide a more comprehensive evaluation of image quality. Recent research has introduced multi-modal prompt-based methodologies for IQA [12]. These approaches use carefully crafted prompts to synergistically mine semantic information from both visual and linguistic data. For instance, a multi-layer prompt structure in the visual branch enhances the adaptability of vision-language (VL) models, while a dual-prompt scheme in the text branch helps the model recognize and differentiate between scene categories and distortion types. Another advancement involves vision-language consistency guided multi-modal prompt learning [12]. This method introduces learnable textual and visual prompts in the language and vision branches of VL models. It also includes a text-to-image alignment quality prediction task, which uses vision-language consistency knowledge to guide the optimization of multi-modal prompts.

Multi-Modal Large Language Models (MLLMs) provide language-based, human-like evaluation of image quality. These models can process and integrate information from multiple modalities, providing a more nuanced assessment of image quality [12]. Further, multi-modal approaches are particularly beneficial for Blind Image Quality Assessment (BIQA), where no reference image is available. By incorporating semantic information from both visual and textual data, these models can achieve higher accuracy and robustness in quality predictions [12].

Multi-modal IQA models benefit significantly from the integration of semantic information, allowing them to treat different types of objects distinctly. This leads to more accurate and reliable assessments of image quality. Further, these models demonstrate competitive performance across various datasets, showcasing their robustness and adaptability. For example, multi-modal prompt-based methodologies have achieved high Spearman Rank Correlation Coefficient (SRCC) values, indicating strong performance in diverse contexts. Multi-modal IQA is applicable in various fields, including photography, medical imaging and autonomous driving. By providing a more comprehensive evaluation of image quality, these models can enhance the performance and reliability of systems that rely on high-quality visuals.

**Fusion Methods:** These methods integrate data from multiple sources to provide a holistic evaluation of the quality of an image and play a crucial role in enhancing the overall assessment. There are many ways a fusion can be carried out. In pixel-level fusion, directly manipulation of pixels of source images is carried out to improve image quality. The pixel-level fusion [13] involves combining pixel values from different modalities in straightforward manner and preserves the spatial details of the original images. Combination may involve weighted averaging where pixel values from different images are averaged based on predefined weights. Principal Component Analysis (PCA) has also been used in performing fusion where transformation of the pixel values is carried out to take them to a new coordinate system, where the principal components are then fused to create a composite image.

In feature-level fusion, features are extracted from the source images and are combined to create a fused image. The fusion focuses on integrating the most relevant information from each modality, enhancing the overall quality and interpretability of the image [14]. There are different approaches adopted for it. For example, in convolutional sparse representation, the technique involves decomposing images into sparse representations and then fusing these representations based on their focus similarity [14]. In wavelet transform-based approach, images are decomposed into different frequency bands, which are then fused to retain both high-frequency details and low-frequency information.

In decision-level fusion, the decisions or outputs from multiple IQA models are combined to provide a final assessment. Decision-level fusion is useful when different models provide complementary information [15]. In a majority of the voting-based decision-level fusion, the final decision is taken based on the majority vote obtained from multiple models. In Bayesian fusion, Bayesian methods combine the probabilistic outputs of different models to make a final decision.

In hybrid fusion methods, combining the information from pixel-level, feature-level and decision-level fusion techniques is carried out to leverage the strengths of each approach. Hybrid methods are designed to provide a more robust and accurate assessment of image quality [15]. Multi-scale decomposition approach of hybrid fusion involves decomposing images at multiple scales and then fusing the decomposed components. It combines the advantages of pixel-level and feature-level fusion. In deep learning-based fusion, deep learning models, such as convolutional neural networks (CNNs), are used to learn and fuse features from different modalities. These models can automatically learn the optimal fusion strategy from the data.

In optimization-based fusion, optimization algorithms are used to find the best fusion strategy that maximizes the quality of the fused image. Optimization-based fusion can be applied at both pixel and feature levels [13]. Particle Swarm Optimization (PSO) has been used to optimize the weights and parameters of the fusion process. Further, the use of Genetic Algorithms (GA) is also being attempted where GA is used to evolve and select the best fusion strategy based on a fitness function that measures image quality.

### 11.2.4  *Personalized image quality assessment*

Personalized image quality assessment is an emerging field. It aims to tailor image quality assessments to individual preferences, viewing conditions and specific application requirements. These types of approaches recognize that different users may have different perceptions of image quality. So, unlike traditional image quality assessment methods, which aim for a one-size-fits-all approach, personalized image quality assessment considers the unique needs and subjective preferences of users. The key considerations while designing personalized image quality assessment techniques include use of user-centric metrics, context-aware assessment and development of adaptive algorithms:

1. **User-Centric Models:** The user-centric models involve developing metrics that align with individual user preferences. They take into account the user preferences and the viewing habits in assessing the quality of images. This can include factors such as color preferences, sharpness and specific artifacts that a user might find more or less tolerable.

2. **Context-Aware Assessment:** The quality of an image can be context-dependent. These techniques adjust the assessment criteria based on the specific context and user requirements for carrying out the quality assessment. For instance, an image used for medical diagnosis requires

different quality criteria compared to an image for social media. Personalized image quality assessment adapts the assessment based on the intended use of the image.

3. **Adaptive Algorithms:** Machine learning and deep learning algorithms play a crucial role in personalized image quality assessment. These algorithms can learn from user feedback to improve their predictions over time, making the assessment more accurate and personalized.

Personalized image quality assessment has got its applications in many domains such as medical imaging, photography and streaming services. In the domain of medical imaging, tailoring image quality to the preferences of radiologists or specific diagnostic requirements has an important. Similarly, in photography, it helps photographers and editors in achieving the desired aesthetic quality based on their unique style. In streaming services, adjusting video quality in real time based on viewer preferences and device capabilities provides improved user experience.

Though personalized image quality assessment provides many advantages, its implementation encounters several challenges. Subjectivity involved in the personalized image quality assessment poses the biggest challenge. Different users have different perceptions of quality, making it challenging to develop universally applicable models. Further, gathering sufficient and relevant data to train personalized models can be difficult in many scenarios. Also, ensuring that personalized assessments can be performed in real time without significant computational overhead is another challenge.

More work in the domain of personalized image quality assessment is required in developing intuitive interfaces for users to provide feedback on image quality, in integration with Augmented Reality (AR) and Virtual Reality (VR) for adapting image quality assessment for immersive experiences, and to achieve cross-platform consistency to ensure the consistent quality assessment across different devices and platforms.

## 11.2.5 *Enhanced metrics and models*

Traditional metrics such as PSNR and SSIM used in image quality assessment have limitations in capturing all aspects of image quality. To overcome these limitations, new metrics and models are being developed to address these limitations. Following are some important developments in this direction:

**Perceptual Metrics:** These metrics better align with human visual perception and provide a more realistic assessment of the image quality. The Visual Information Fidelity (VIF) and the Learned Perceptual Image Patch Similarity (LPIPS) metrics are examples of perceptual metrics.

Visual Information Fidelity (VIF) is an advanced metric used in Image Quality Assessment (IQA) to evaluate the quality of an image based on the amount of visual information it retains compared to a reference image. Unlike traditional metrics such as Peak Signal-to-Noise Ratio (PSNR) and Structural Similarity Index (SSIM), which primarily focus on pixel-level differences, VIF aims to measure the fidelity of the visual information that is perceived by the human visual system. VIF is grounded in information theory, which quantifies the amount of information present in an image. It assesses how much of the original visual information is preserved in the distorted image. It leverages natural scene statistics to model the human visual system's response to images. This helps in capturing the perceptual quality of images more accurately. Further, it performs a multi-scale analysis, evaluating the image at different resolutions and scales, allowing the metric to capture both fine details and overall structural information. By focusing on the information that is perceptually relevant to human observers, VIF provides a more accurate assessment of image quality from a human perspective. It is robust to various types of distortions, including noise, blur and compression artifacts. The computation of VIF involves mainly three steps, *viz.* image decomposition, information fidelity measurement and aggregation. In decomposition, both the reference and distorted images are decomposed into multiple scales using wavelet transforms. This process separates the image into different frequency bands, capturing various levels of detail. The next step is information fidelity measurement where at each scale, the amount of visual information in the distorted image is compared to the reference image. This is done by measuring the mutual information between the corresponding wavelet coefficients of the two images. In the aggregation step, the information fidelity measurements from all scales are aggregated to produce the final VIF score. This score represents the overall fidelity of the visual information in the distorted image.

Learned Perceptual Image Patch Similarity (LPIPS) is another advanced metric used in Image Quality Assessment (IQA) to evaluate the perceptual similarity between two images. Unlike traditional metrics that rely on pixel-level differences, LPIPS leverages deep learning to measure the similarity based on the activations of neural network layers, which better align with human visual perception.

Key concepts of LPIPS include deep features, perceptual similarity and network types. LPIPS uses deep features extracted from pre-trained deep neural networks, such as AlexNet, VGG and SqueezeNet. These features capture high-level information about the image content, such as textures and structures, which are more relevant to human perception. Further, to measure the perceptual similarity, this metric computes the distance between the deep features of two image patches. A lower LPIPS score indicates higher perceptual similarity, meaning the images look more alike to human observers. The use of a network plays an important role in assessing image quality as each network captures different aspects of perceptual similarity. The value LPIPS metric can be computed using different backbone deep networks, such as AlexNet, VGG and SqueezeNet.

The computation of LPIPS involves primarily three steps: feature extraction, distance measurement and aggregation. In feature extraction, both the reference and distorted images are passed through a pre-trained neural network to extract deep features from multiple layers. Subsequently, in distance measurement, the Euclidean distance between the corresponding deep features of the two images is computed. This distance reflects the perceptual dissimilarity between the images. In the aggregation step, the distances from all layers are aggregated to produce the final LPIPS score. This score represents the overall perceptual similarity between the images.

As stated above, the LPIPS metric is aligned with human perception and is correlated well with human understanding of image quality and hence makes it a reliable metric for perceptual similarity. It is also robust to various types of distortions, such as noise, blur and compression artifacts. Further, LPIPS provides flexibility as it can be computed considering different backbone networks, allowing capturing of different aspects of image quality.

**Hybrid Models:** Hybrid models combine the strengths of traditional metrics and advanced machine learning techniques to provide a more comprehensive and accurate evaluation of image quality. These models leverage both handcrafted features and learned features from deep neural networks, aiming to capture a wide range of image distortions and perceptual qualities. By combining traditional and deep learning features, hybrid models provide a more accurate assessment of image quality that aligns closely with human perception. These models are robust to a wide range of distortions, including noise, blur, compression artifacts and more. Further, hybrid models can be tailored to specific applications and types of images, making them versatile tools for image quality assessment.

Hybrid models rely on combinations of features or multi-stage processing for quality estimation. These models integrate features from traditional image quality assessment metrics, such as PSNR and SSIM, with deep features extracted from a neural network. This combination allows the model to capture both low-level pixel information and high-level perceptual information. These models often involve multiple stages of processing, where initial assessments are carried out using traditional metrics and afterward, the obtained results are refined with the help of deep learning models. This multi-stage approach enhances the robustness and accuracy of the assessment. Hybrid models can also adaptively weight different features based on their relevance to the specific type of distortion or image content. This flexibility improves the model's ability to handle diverse image quality issues.

For example, in a hybrid model of image quality assessment, first traditional metrics such as PSNR and SSIM are used to perform an initial assessment of the image quality. Further, deep features are extracted from the image using a pre-trained neural networks, such as VGG and ResNet. Subsequently, the traditional and deep features are combined using adaptive weighting to form a comprehensive feature set. Further, a machine learning model such as a regression model or a neural network model is used on the combined features to predict the final image quality score.

### 11.2.6   *Applications in emerging technologies*

Image quality assessment plays a crucial role in various emerging technologies, ensuring that visual data meets the high standards required for these advanced applications. Following are some key areas where image quality assessment has shown a significant impact:

1. **Virtual Reality (VR) and Augmented Reality (AR):** These technologies rely heavily on high-quality visual content to provide immersive and interactive experiences. IQA is essential in these fields to ensure that the images and videos are clear, detailed and free from distortions. In VR applications, the quality of the visual content directly affects the user's sense of presence and immersion. High-quality images and videos are crucial for applications, such as gaming, virtual tours and training simulations. Further, in AR Applications, digital content is

overlaid on the real world. Image quality assessment is necessary here to ensure that this content is of high quality required for applications like navigation, education and industrial maintenance.

2. **Autonomous Vehicles:** Autonomous vehicles rely on a variety of sensors, including cameras, to navigate and make decisions. The quality of the visual data captured by these cameras is critical for the safety and reliability of these vehicles. Object detection is a crucial part of an autonomous vehicle and high-quality images are necessary for accurate object detection and recognition, which are very crucial for avoiding obstacles and ensuring safe navigation of the vehicle. Another task that is very important in autonomous vehicles is about acquiring correct environmental perception. Clear and detailed images help autonomous vehicles perceive their environment accurately, enabling them to make informed decisions in real time.

3. **Medical Imaging:** Medical imaging technologies, such as MRI, CT scans, and X-rays, require high-quality images for accurate diagnosis and treatment planning. IQA ensures that these images are clear and detailed, allowing healthcare professionals to make precise assessments. For example, high-quality medical images are essential for detecting and diagnosing various conditions, from fractures to tumors. Similarly, clear and detailed images help in planning surgical procedures and other treatments.

4. **Remote Sensing and Satellite Imaging:** Remote sensing and satellite imaging are used for a variety of applications, including environmental monitoring, urban planning and disaster management. IQA ensures that the images captured by satellites and drones are of high quality, providing accurate and reliable data. For example, high-quality images are essential for monitoring changes in the environment, such as deforestation, urbanization and climate change. Similarly, clear and detailed images help in assessing the impact of natural disasters and planning effective response strategies.

5. **Smart Cities:** Smart cities leverage advanced technologies to improve urban living. High-quality visual data are crucial for various smart city applications, including traffic management, surveillance and infrastructure monitoring. For example, good-quality images from traffic cameras help in monitoring and managing traffic flow, reducing congestion and improving safety. Moreover, clear and detailed images are essential for effective surveillance, helping in crime prevention and ensuring public safety.

6. **E-commerce and Online Retail:** In e-commerce and online retail, high-quality images are crucial for showcasing products and enhancing the shopping experience. IQA ensures that product images are clear, detailed and visually appealing. For example, high-quality images help in presenting products accurately, allowing customers to make informed purchasing decisions. Further, clear and detailed images enhance the overall shopping experience, leading to higher customer satisfaction and reduced return rates.

### 11.2.7  *Standardization and benchmarking*

Standardization and benchmarking are critical components in the field of Image Quality Assessment (IQA). They ensure consistency, reliability and comparability of IQA methods and results across different applications and industries. A detailed look at how these processes work and their importance in IQA is elaborated in the following.

Standardization in IQA involves establishing a set of standards or guidelines that define the processes, metrics and practices used in IQA. These standards ensure that IQA methods are applied consistently, making the results reliable and comparable. Organizations and industry bodies develop standards for IQA to ensure uniformity. These standards specify the metrics to be used, the conditions under which assessments should be conducted and the procedures for evaluating image quality. Further, standardization ensures that IQA methods produce consistent and reliable results, regardless of the specific application or context. This is crucial for comparing image quality across different systems and technologies. Moreover, by adhering to standardized methods, different systems and devices can work together seamlessly. This is particularly important in fields like medical imaging, where images from different devices need to be compared and analyzed consistently.

Benchmarking in IQA involves comparing the performance of different IQA methods, systems or processes against established standards or best practices. It helps identify areas for improvement and drives innovation. Benchmarking allows organizations to compare their IQA methods against industry standards or the performance of leading systems. This helps identify strengths and weaknesses and guides improvements. By benchmarking against the best practices in the industry, organizations can adopt proven strategies and techniques to enhance their IQA processes. This leads to better image quality and more accurate assessments. Benchmarking is an ongoing process that encourages continuous improvement. By regularly comparing performance against benchmarks, organizations can stay ahead of industry trends and maintain high standards of image quality.

Examples of standardization and benchmarking in IQA include standardized metrics and benchmarking frameworks. Metrics such as Peak Signal-to-Noise Ratio (PSNR) and Structural Similarity Index (SSIM) are widely used in IQA due to their standardized definitions and calculation methods. These metrics provide a common basis for comparing image quality across different systems. Organizations such as the Video Quality Experts Group (VQEG) develop benchmarking frameworks that provide standardized evaluation protocols and datasets for comparing different IQA methods. These frameworks ensure that comparisons are fair and consistent.

# References

[1] D. M. Chandler, "Seven challenges in image quality assessment: Past, present, and future research," *International Scholarly Research Notices*, 2013(1), 905685, 2013. Available: https://onlinelibrary.wiley.com/doi/abs/10.1155/2013/905685

[2] L. Leveque, J. Yang, X. Yang, P. Guo, K. Dasalla, L. Li, Y. Wu, and H. Liu, "CUID: A new study of perceived image quality and its subjective assessment," in *Proc. of IEEE International Conference on Image Processing (ICIP 2020)*, 2020, pp. 116–120.

[3] S. Athar and Z. Wang, "A comprehensive performance evaluation of image quality assessment algorithms," *IEEE Access*, 7, 140030–140070, 2019.

[4] R. Liao, C. Hui, Y. Lang, and F. Jiang, "S-IQA: Image quality assessment with compressive sampling," 2024. Available: https://arxiv.org/html/2404.17170v1

[5] C. Wang, X. Lv, W. Ding, and X. Fan, "No-reference image quality assessment with multi-scale weighted residuals and channel attention mechanism," p. 13449–13465, 2022.

[6] S. A. Golestaneh, S. Dadsetan, and K. M. Kitani, "No-reference image quality assessment via transformers, relative ranking, and self-consistency," 2022. Available: https://arxiv.org/abs/2108.06858

[7] S. Gu, J. Bao, D. Chen, and F. Wen, "GIQA: Generated image quality assessment," in *Proceedings of European Conference on Computer Vision (ECCV 2020)*, 2020, pp. 369–385.

[8] Z. A. Chami, C. A. Jaoude, R. Chbeir, M. Barhamgi, and M. N. Alraja, "A no-reference and full-reference image quality assessment and enhance-

ment framework in real-time," *Multimed Tools and Applications*, p. 32491–32517, 2022.

[9] "Hardware acceleration of most apparent distortion image quality assessment algorithm on fpga using opencl," Master Thesis, Arizona State University, 2017.

[10] J. Holloway, V. Kannan, Y. Z. 3ORCID, D. M. Chandler, and S. Sohoni, "GPU acceleration of the most apparent distortion image quality assessment algorithm," *Journal of Imaging*, 4(10), 111, 2017.

[11] T. D. Phan, S. Sohoni, D. M. Chandler, and E. C. Larson, "Performance-analysis-based acceleration of image quality assessment," in *Proc. of IEEE Southwest Symposium on Image Analysis and Interpretation, SSIAI 2012*, 2012, pp. 81–84. [Online]. Available: https://doi.org/10.1109/SSIAI.2012.6202458

[12] W. Pan, T. Gao, Y. Zhang, R. Hu, X. Zheng, E. Zhang, Y. Gao, Y. Liu, Y. Shen, K. Li, S. Zhang, L. Cao, and R. Ji, "Multi-modal prompt learning on blind image quality assessment," 2024. Available: https://arxiv.org/abs/2404.14949

[13] S. Basu, S. Singhal, and D. Singh, "A systematic literature review on multimodal medical image fusion," *Multimed Tools and Applications*, 83, 15845–15913, 2024. [Online]. Available: https://doi.org/10.1007/s11042-023-15913-w

[14] Y. Hu, P. Wu, B. Zhang, W. Sun, Y. Gao, C. Hao, and X. Chen, "A new multi-focus image fusion quality assessment method with convolutional sparse representation," *The Visual Computer*, 2024. Available: https://doi.org/10.1007/s00371-024-03351-0

[15] J. Zheng, J. Xiao, Y. Wang, and X. Zhang, "CIRF: Coupled image reconstruction and fusion strategy for deep learning based multi-modal image fusion," *Sensors*, 24(11), 2024. [Available: https://doi.org/10.3390/s24113545

# Chapter 12

# Emerging Trends and Future Directions

This chapter delves into the exploration of emerging trends and future directions in image quality assessment (IQA). It serves as a compass, guiding readers through the ever-evolving landscape of image quality assessment, from current trends to promising future paths.

## 12.1 Emerging trends in Image Quality Assessment

The field of image quality assessment is constantly evolving, with researchers exploring novel metrics and innovative approaches. Among the emerging trends, there is a focus on developing sophisticated models capable of handling diverse image types and challenging conditions. These advancements hold the promise of significantly enhancing image quality across various industries.

### 12.1.1  *Blind image quality assessment*

Blind Image Quality Assessment (BIQA) is the task of evaluating image quality without relying on the original reference image. This challenging endeavor is particularly relevant in real-world scenarios where reference images may not be available. Recent research has witnessed significant advancements in BIQA, spanning both hand-crafted methods — tailored for specific distortions — and deep-learned approaches that leverage supervised and unsupervised learning techniques [1, 2].

### 12.1.2   *Multi-modal quality assessment*

Traditional image quality assessment methods predominantly focus on visual information alone. However, human perception is a multi-sensory experience, integrating various modalities, such as sight, sound and text. Multi-modal blind image quality assessment seeks to emulate our brain's ability to process these diverse modalities simultaneously. By considering multiple sensory cues, this approach enhances the accuracy of quality assessment, bridging the gap between objective metrics and human perceptual judgments [1].

### 12.1.3   *Data-driven approaches*

Data-driven no-reference image quality assessment models, leveraging convolutional neural networks (CNNs), have indeed outperformed hand-crafted features. These models benefit from techniques, such as patch-wise training, transfer learning and quality-aware pre-training. These strategies effectively address the challenge of limited human-labeled data, enabling more accurate and robust quality assessment [3].

### 12.1.4   *Patch-wise training and transfer learning*

Researchers have indeed explored patch-wise training approaches, enabling models to focus on local image regions. By doing so, they capture fine-grained quality variations, leading to improved overall assessment accuracy. Additionally, transfer learning has emerged as a powerful technique. It involves leveraging pre-trained convolutional neural networks (CNNs) with weights learned from large-scale image classification tasks. Through transfer learning, image quality assessment models can tap into features acquired from diverse visual data, enhancing their performance.

### 12.1.5   *Quality-aware pre-training*

In recent advancements, researchers have introduced quality-aware pre-training techniques for image quality assessment. These models undergo pre-training where they learn to predict image quality scores. By incorporating quality-awareness during this initial phase, the models develop a deeper understanding of image nuances, artifacts, and perceptual features. Consequently, when applied to actual quality assessment tasks, they exhibit enhanced accuracy and robustness. Quality-aware pre-training bridges the gap between low-level visual features and high-level perceptual quality, contributing to more effective image quality evaluation.

### 12.1.6  *Unified vision-language pre-training for quality and aesthetics (UniQA)*

This approach recognizes that human perception of visual content involves more than just technical quality [4]. Aesthetic appeal plays a crucial role in how we perceive images. By unifying quality assessment and aesthetic evaluation, the model aims to simulate the holistic subjective experience that humans have when viewing images. There are different components which are involved in it. These are explained in the following:

1. **Image Quality Assessment (IQA):** IQA focuses on technical aspects, such as sharpness, noise and artifacts. Traditional IQA models often rely on hand-crafted features or deep learning approaches. In this unified approach, IQA is seamlessly integrated into the vision-language framework.
2. **Image Aesthetic Assessment (IAA):** IAA considers subjective factors related to beauty, composition and emotional impact. Aesthetic assessment is inherently challenging due to its subjectivity. By combining IAA with IQA, the model aims to capture both technical and aesthetic dimensions.
3. **Vision-Language Pre-Training:** The model learns joint representations of images and text (captions, descriptions, etc.). Pre-training involves training on large-scale vision-language datasets. During this phase, the model learns to associate visual features with textual context.
4. **Fine-Tuning for Specific Tasks:** After pre-training, the model can be fine-tuned for specific downstream tasks. For quality and aesthetics, fine-tuning involves using annotated data to refine the joint representations. The model adapts to specific quality and aesthetic cues relevant to the task.

By considering both quality and aesthetics, the model mimics how humans perceive images. Further, joint pre-training enhances robustness by leveraging shared features. The technique has got its applications in many domains. For example, content creators can benefit from automated assessments of both quality and aesthetics. Other applications include photo curation, social media and creative content generation. This unified vision-language approach bridges the gap between technical quality and artistic appeal, enriching our understanding of visual content.

### 12.1.7    *Deep feature statistics*

Traditional image quality assessment often relies on human ratings or subjective opinions. Collecting human judgments can be time-consuming, expensive and inherently subjective due to individual preferences and biases. Opinion-Unaware Blind Image Quality Assessment (OU-BIQA) [5] emerges as a solution to this challenge by aiming to assess image quality without explicitly relying on human opinions. Different components of OU-BIQA are as follows:

1. **Deep Features from Pre-Trained Visual Models:** OU-BIQA models leverage deep features extracted from pre-trained convolutional neural networks (CNNs). These features capture high-level representations of image content, including texture, edges, and semantic information. By using pre-trained models, OU-BIQA benefits from the rich knowledge acquired during large-scale visual tasks (e.g., image classification).
2. **Statistical Analysis:** Instead of relying on human ratings, OU-BIQA models perform statistical analysis on the extracted deep features. This analysis considers feature distributions, correlations and other statistical properties. By quantifying these properties, the model infers image quality, effectively bridging the gap between objective metrics and human perception.

The OU-BIQA offers several advantages. It avoids the need for explicit human opinions, reducing subjectivity and potential biases. Further, the model's predictions are based on data-driven evidence rather than individual preferences. By bypassing the collection of human rating data, OU-BIQA improves training efficiency. This is especially valuable when dealing with large datasets or real-time applications.

Though the OU-BIQA offers several advantages, there are some challenges too. Ensuring that statistical features generalize well across diverse image datasets is crucial. Further, robustness across different image domains remains an active area of research. Further, designing effective statistical measures that capture various quality aspects (e.g., sharpness, noise and artifacts) is challenging. Researchers explore novel statistical techniques to enhance assessment accuracy.

OU-BIQA finds applications in various domains, such as image compression, image enhancement and quality control (for example, assessing product images in manufacturing). Future research may focus on refining statistical models, addressing domain-specific challenges, and integrating multi-modal cues (e.g., text and audio) for comprehensive quality assessment. In summary, opinion-unaware blind image quality assessment models combine data-driven features and statistical analysis, providing an objective and efficient approach to evaluate image quality.

### 12.1.8  *Large-scale datasets*

Image quality assessment (IQA) is pivotal in various applications, influencing the selection of high-quality images and guiding compression and enhancement methods. Researchers recognize the importance of robust IQA models and have made significant efforts to create large-scale datasets for training and evaluation. One notable dataset is the DepictQA dataset, introduced in the work by You *et al.* [6]. This dataset comprises a staggering 495,000 images, making it a valuable resource for IQA research. By including diverse images across various domains, DepictQA enhances both data quality and representation. Researchers can leverage this extensive dataset to develop and validate IQA models that generalize well to real-world scenarios. The availability of such large-scale datasets empowers the IQA community to advance the field, ensuring better image quality assessment across applications

### 12.1.9  *Use of generative models*

Recent trends in image quality assessment include leveraging generative models to learn low-dimensional representations using autoencoding. These models preserve better visual features and improve generalization across datasets. One such method is VAE-QA [7], which efficiently predicts image quality in the presence of a full reference.

Autoencoders have gained prominence in learning compact and meaningful representations of data. These models consist of an encoder that maps input data (such as images) to a low-dimensional latent space, and a decoder that reconstructs the original data from this representation. By capturing essential features in this reduced dimensionality, autoencoders help preserve visual information. The Variational Auto-Encoder for Quality Assessment (VAE-QA) [7] is a specific method that combines generative modeling with quality assessment. It efficiently predicts image quality while considering a full-reference (FR) image as a reference. It involves two steps:

1. **Encoding:** VAE-QA encodes an input image into a low-dimensional latent representation.
2. **Quality Prediction:** It then predicts the quality of the encoded image using this representation.
3. **Decoder and Reconstruction:** The decoder reconstructs the image from the latent space, allowing for quality assessment. VAE-QA benefits from both the generative power of autoencoders and the context provided by the full-reference image.

The VAE-QA offers several advantages. By learning meaningful features, VAE-QA generalizes better across diverse datasets. It efficiently predicts quality without relying solely on hand-crafted features or extensive human ratings. The VAE-QA implementation also faces certain challenges and that leads to future research. For example, we can explore how different regions of the latent space correspond to specific quality attributes. Further, integration of other modalities (e.g., text and audio) into the VAE-QA framework can be done for comprehensive assessment. In totality, leveraging generative models such as VAE-QA opens exciting avenues for accurate and efficient image quality assessment, benefiting various applications in computer vision and beyond.

### 12.1.10   *Explainable IQA*

Explainable Image Quality Assessment (x-IQA) is an essential aspect, particularly for medical imaging. This is because explainable image quality assessment enhances our understanding of image quality issues, leading to better medical diagnoses. Further, in medical imaging, the quality of acquired images significantly impacts diagnosis accuracy. Poor-quality images can lead to misdiagnosis or incorrect treatment decisions. Ensuring high-quality images is crucial for patient care. Researchers in this domain have explored automated methods to assess image quality. However, explaining these methodologies remains an underexplored area. Researchers have proposed an explainable system to evaluate image quality. They validate this idea using two objectives: Foreign Object Detection on Chest X-Rays (Object-CXR) and Left Ventricular Outflow Tract (LVOT) Detection on Cardiac Magnetic Resonance (CMR) Volumes. The system relies on NormGrad, an algorithm that efficiently localizes image quality issues using saliency maps from a classifier. There are various techniques to measure the faithfulness of saliency detectors. However, NormGrad outperforms other saliency detection methods. It achieves a repeated Pointing Game score of 0.853 for Object-CXR and 0.611 for LVOT datasets [8].

In another work presented in Ref. [9], large language models have been used to generate the quality of an image equivalent to human-readable assessments. The work introduces a novel approach for evaluating text-to-image generation methods using visual large language models (LLMs). It utilizes a hierarchical Chain of Thought (CoT) with MiniGPT-4 to produce self-consistent, unbiased texts. The technique is found to be cost-effective and efficient compared to human evaluation. It is able to distinguish between real and generated images. It evaluates text-image alignment and image aesthetics and requires no model training or fine-tuning. It enhances transparency and explainability. It demonstrates similar performance to state-of-the-art methods on COCO Caption and overcomes limitations in handling ambiguous prompts and text recognition in generated images. An interesting survey of techniques is presented in Ref. [10] to understand a broader perspective on explainability in large language models.

### 12.1.11  *Unified IQA models*

Unified Image Quality Assessment (IQA) models aim to integrate various types of IQA into a single framework. This approach can streamline the evaluation process by combining different IQA methods, such as No-Reference (NR), Full-Reference (FR) and Reduced-Reference (RR) assessments, into one cohesive model. Recent research has shown promising results in this area. For instance, a comprehensive collection of IQA papers and resources on GitHub highlights efforts to unify all IQA types into a single model. Additionally, a computational framework has been proposed to integrate model-centric and data-centric IQA approaches, enhancing the overall performance and generalization of IQA models [11]. These unified models leverage advanced techniques such as deep learning and multi-layer feature extraction to assess image quality more accurately and efficiently [12]. By integrating different IQA types, these models can provide a more holistic evaluation of image quality, which is crucial for various applications in image processing and computer vision.

## 12.1.12    *Quality estimation of AI-generated images*

Assessing the quality of AI-generated images (AIGIs) is crucial for evaluating their performance and guiding their generation. Here are some notable works in this field:

1. **Generated Image Quality Assessment (GIQA):** Traditional image quality assessment methods are not well suited for generated images due to their unique artifacts and variations. Generated image quality assessment is a field of research that focuses on evaluating the quality of images produced by generative models, such as deep neural networks. The goal is to develop metrics or methods that can objectively measure the quality of synthetically generated images. The work presented in Ref. [13] introduces a novel approach to evaluating the quality of images generated by Generative Adversarial Networks (GANs). It proposes three GIQA algorithms which are categorized into learning-based and data-based approaches. These algorithms are designed to evaluate the quality of generated images in a way that aligns with human visual perception. The work in Ref. [13] is evaluated on images generated by various recent GAN models across different datasets. Another work in this direction is proposed in Ref. [14].

    GIQA framework represents a significant advancement in the field of image quality assessment, particularly for images generated by GANs. By providing a method to objectively and automatically assess the quality of each generated image, GIQA has the potential to enhance the development and evaluation of generative models.

2. **SF-IQA—Quality and Similarity Integration for AI-Generated Image Quality Assessment:** This work introduces SF-IQA [15], a novel image quality metric specifically designed for AI-Generated Images (AIGIs). SF-IQA integrates quality and similarity in a score fusion manner. It uniquely combines image quality and text-image similarity, addressing both perceptual quality and semantic consistency between AIGIs and their corresponding textual prompts. It employs a sophisticated multi-layer feature extractor and fusion module to extract quality-aware features. This process aggregates local and global-level features, allowing for the extraction of quality-aware features. Further, it fine-tunes a vision-language model for image-text similarity alignment.

SF-IQA achieved state-of-the-art results on AGIQA-3K benchmarks and ranked fourth in the NTIRE 2024 Quality Assessment of AI-Generated Content Challenge [16].

3. **Bridging the Synthetic-to-Authentic Gap—Distortion-Guided Unsupervised Domain Adaptation for Blind Image Quality Assessment:** This work proposes a novel approach to address the challenge of domain adaptation between synthetic and real-world images for Blind Image Quality Assessment (BIQA) [17]. The BIQA involves assessing image quality without access to reference images. Training models on synthetic data can be beneficial, but they often suffer from poor generalization to real-world domains due to domain gaps. The proposed framework, Distortion-Guided Unsupervised Domain Adaptation (DGQA), leverages adaptive multi-domain selection using prior knowledge from distortion. It aims to match the data distribution between source (synthetic) and target (real-world) domains, reducing negative transfer from outlier source domains. Extensive experiments demonstrate the effectiveness of DGQA in two cross-domain settings: synthetic distortion to authentic distortion and synthetic distortion to algorithmic distortion. DGQA is orthogonal to existing model-based BIQA methods and can improve performance with less training data

4. **AIGCOIQA2024—Perceptual Quality Assessment of AI-Generated Omnidirectional Images:** The authors in this work address the lack of dedicated Image Quality Assessment (IQA) criteria for AI-generated omnidirectional images. The work establishes a large-scale AI-generated omnidirectional image IQA database, known as AIG-COIQA2024 [18], containing 300 omnidirectional images generated from 5 AI-generated content (AIGC) models using 25 text prompts. Human visual preferences are assessed from three perspectives, *viz.* quality, comfortability and correspondence, and corresponding human preference ratings are generated. The database aims to facilitate future research in this domain and understand human visual preferences for AI-generated omnidirectional images in relation to the development of image quality assessment techniques for such images. The earlier similar work, known as AIGCIQA2023 [19], provides a better understanding of human visual preferences for AIGIs based on text prompts and includes generated images from six deep generative models and corresponding subjective quality ratings.

## 12.2    Future directions

In this comprehensive book, we delve into the fundamental concepts of image quality. As readers explore its pages, they gain valuable insights into a diverse array of image quality assessment techniques and quality-aware enhancement methods. From understanding the intricacies of perceptual quality metrics to exploring cutting-edge algorithms, this book equips researchers and practitioners with the knowledge needed to navigate the dynamic field of image quality. Additionally, we illuminate potential future directions for research, inviting readers to contribute to the ongoing evolution of this critical domain:

1. **Deep Learning for No-Reference Image Quality Assessment and Enhancement:** Deep learning has become a cornerstone in both full-reference (FR) and reduced-reference (RR) image quality assessment. Its ability to learn intricate features from data has significantly improved assessment accuracy. However, an exciting avenue lies in the development of versatile, no-reference-based techniques. These methods would assess image quality without relying on a reference image, making them applicable in real-world scenarios where references might be absent. By harnessing deep learning, we can create models that not only evaluate image quality but also enhance it—whether by denoising, sharpening or addressing other imperfections. This convergence of deep learning and no-reference assessment holds great promise for advancing image quality evaluation.

2. **Addressing of Unexplored Distortions:** In this book, novel no-reference and no-training-based quality assessment and enhancement techniques have been introduced for various image distortions. These include poor contrast, illumination issues and noise. However, certain distortions—such as ringing artifacts in JP2K, fast fading and blur—have received less attention in the context of training-free NR-image quality assessment. A promising future direction would involve developing distortion-specific NR assessment techniques tailored to these unexplored distortions.

3. **Handling of Multiple Distortions:** The development of general-purpose training-free NR-image quality assessment is more complex than general-purpose FR-image quality assessment and RR-image quality assessment due to the unavailability of reference images and training samples. However, general-purpose training-free NR-image quality

assessment is useful for applications with no reference image and training data and needs to handle multiple distortions. An attempt is required to develop a unified no-reference and no-training-based technique for quality assessment and enhancement which can handle multiple distortions together.

4. **3D No-Reference Image Quality Assessment:** The designing of reliable three-dimensional (3D) RR-image quality assessment metrics is the future direction of FR-image quality assessment, RR-image quality assessment and NR-image quality assessment. It is challenging due to the exact computations of 3D features in the 3D domain. Developing a training-free NR-image quality assessment technique for 3D data would be a significant impact on research.

    We can understand the need for image quality assessment techniques for 3D data as follows. Three-dimensional data captured by a 3D camera may be distorted due to poor illumination, noise, the different distances between objects and 3D camera and reflectance due to shiny objects. All these factors are responsible for point density variation and distortion of the structure in the captured 3D data. A few examples of 3D data are shown in Figure 12.1. These 3D scenes are captured by a Realsense 3D camera D435 for the purpose of 3D object recognition [20]. The accuracy of 3D object recognition is mainly dependent on the quality of captured data. Image quality assessment technique data can be useful here to reject or control poor-quality 3D data. In addition, 3D quality-aware enhancement techniques can be built to enhance 3D data suitably.

5. **Quantum Image Processing:** Quantum computing is a rapidly advancing field, with researchers and companies worldwide making significant progress in various areas. A notable breakthrough came in 2019 when Google's Sycamore processor achieved "quantum supremacy," demonstrating the potential computational power of quantum systems [21]. IBM is also making major strides, recently unveiling its 1121-qubit superconducting processor, IBM Condor [22], alongside new systems such as IBM Quantum System Two. The company continues to improve qubit coherence and reduce error rates [23–25]. Additionally, the quantum computing landscape is expanding with a growing number of startups bringing innovative technologies and approaches.

    Quantum computing, with its ability to perform complex calculations at exponentially faster speeds than classical computers [26], opens up new possibilities for addressing challenging problems in machine learning. The integration of quantum computing and machine learning [27]

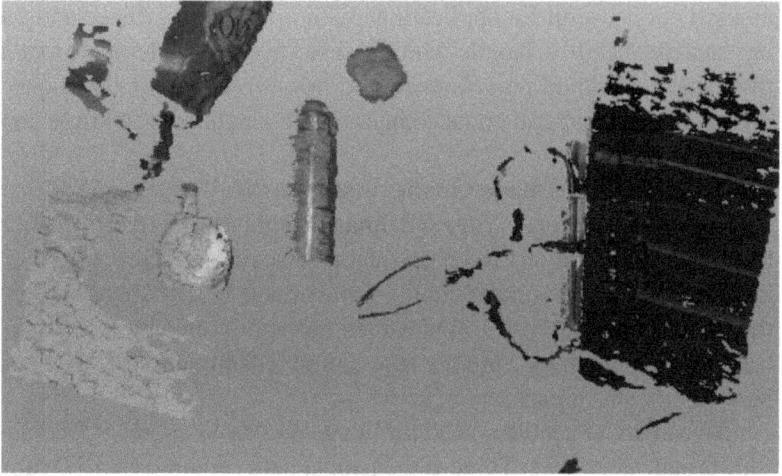

(a)

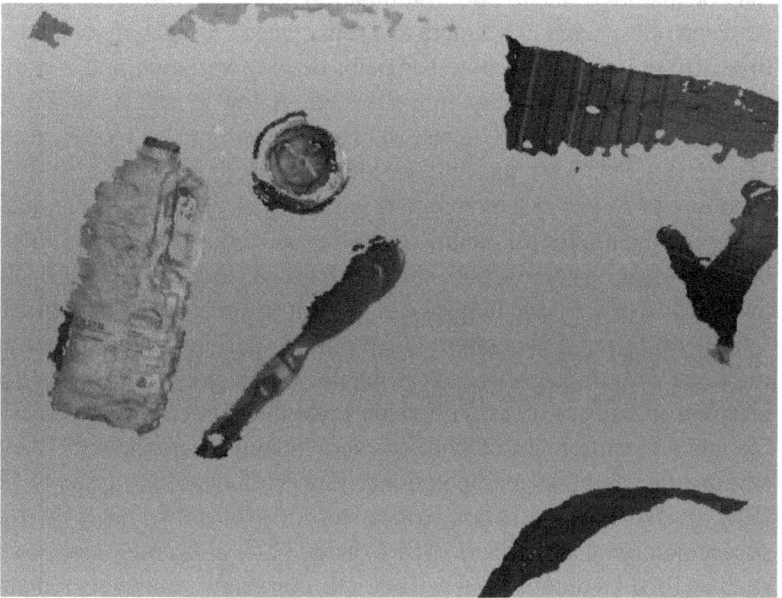

(b)

**Figure 12.1.** A few examples of 3D scenes captured by a Realsense 3D camera D435 [20].

marks a significant shift in information processing and analysis. Quantum machine learning harnesses the distinctive properties of quantum systems, such as superposition and entanglement [28], to improve the efficiency of algorithms used in data analysis [29], pattern recognition [30–35] and optimization [36, 37]. These advancements enable the development of more advanced machine learning models, capable of processing vast amounts of data and solving intricate computations with unparalleled speed [38].

Quantum Image Processing (QIP) combines quantum mechanics with image processing, offering new approaches to managing visual data. As traditional computing faces challenges with the growing complexity of image-related tasks, quantum computing leverages properties such as superposition and entanglement to process information in more innovative and efficient ways. QIP aims to develop algorithms using qubits for tasks such as image compression, denoising, quality assessment and recognition. Figure 12.2 shows the pipeline of quantum image processing. The steps in quantum image processing involve first converting the image from a classical state to a quantum state. The image is then processed in the quantum state, and finally, the processed image is converted back from the quantum state to a classical state as the output.

Hybrid quantum-classical approaches [39–45] represent a transformative paradigm in quantum computing, aiming to combine the strengths of both classical and quantum systems to address complex computational challenges [46, 47]. One such approach that has gained

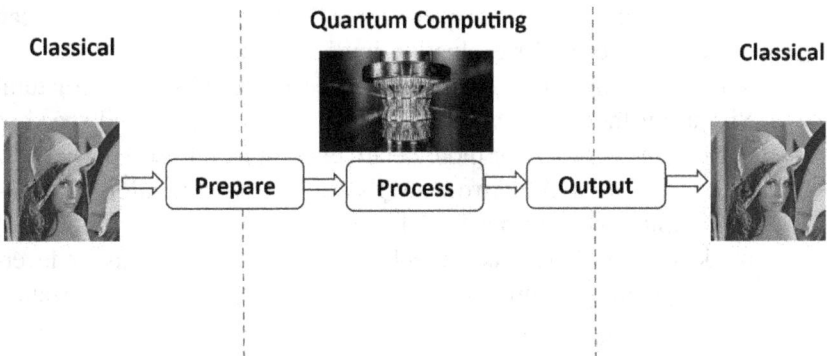

**Figure 12.2.**   Quantum Image Processing.

traction in recent years is hybrid quantum-classical neural networks (HQCNs) [48]. In HQCNs, traditional neural networks, typically composed of classical neuron layers, are integrated with quantum elements, often in the form of variational quantum circuits [49]. The training process adjusts both classical and quantum parameters, employing a hybrid optimization strategy [50]. The classical component manages tasks best suited for classical computation, such as data preprocessing and certain aspects of feature extraction, while the quantum component focuses on quantum-friendly computations, including optimization tasks or utilizing quantum interference for improved pattern recognition.

One of the most widely used traditional techniques for image recognition is the Convolutional Neural Network (CNN). Since its introduction by LeCun [51], CNNs have become a dominant method in numerous machine learning applications. In 2020, Henderson *et al.* [52] extended the principles of CNNs into the quantum domain, giving rise to the Quanvolutional Neural Network (QNN)—a hybrid quantum-classical model that opens up new possibilities for feature extraction and classification in quantum computing.

Here are some ways quantum computing can enhance image quality assessment:

(a) **Quantum Feature Extraction:** Quantum algorithms can process large image datasets in parallel, enabling faster and more effective feature extraction, such as edge detection, color gradients and texture analysis, which are crucial for assessing image quality.

(b) **Quantum Distance Metrics:** Quantum-based algorithms can be used to compute similarity measures between images, which can serve as a metric for evaluating image degradation. Quantum versions of metrics such as Mean Squared Error (MSE) or Structural Similarity Index (SSIM) could improve the accuracy and speed of these comparisons. For more detail, explore Ref. [53].

(c) **Quantum Neural Networks (QNNs):** The combination of quantum computing and machine learning can lead to more powerful models for IQA. QNNs can enhance traditional methods by leveraging quantum computing's ability to solve complex optimization problems, making it easier to assess image distortions, such as noise, blur and compression artifacts.

(d) **Speed and Parallelism:** Quantum computers, with their ability to perform many calculations simultaneously, can accelerate processes in real-time IQA applications. This is particularly beneficial for video quality assessment or other time-sensitive applications, where quick and accurate results are essential.

(e) **Hybrid Quantum-Classical Models:** Hybrid approaches that combine quantum circuits with classical image processing techniques, such as convolutional neural networks (CNNs), can be applied to enhance feature extraction and improve the performance of image quality models.

Here are some ways quantum computing is being explored for image enhancement:

(a) **Quantum Fourier Transforms (QFT):** Quantum algorithms can utilize QFT for frequency domain analysis, enabling faster and more efficient filtering techniques. This helps in removing noise, enhancing image resolution and improving contrast in images.

(b) **Quantum Edge Detection:** Algorithms based on quantum parallelism can significantly improve the edge detection process, an essential component of image enhancement. Quantum Sobel or Prewitt filters can be used for better boundary detection and sharper image details.

(c) **Quantum Image Denoising:** Quantum algorithms are capable of more efficiently detecting and eliminating noise in images. By using quantum versions of techniques such as wavelet transforms, noise can be reduced without losing essential details, which is often a challenge in classical approaches.

(d) **Quantum Super-Resolution:** Quantum computing has the potential to achieve super-resolution in images by reconstructing higher-resolution images from lower-resolution inputs. This is done by using quantum machine learning and optimization algorithms to enhance image details beyond traditional limits.

(e) **Quantum-assisted Image Enhancement Algorithms:** Some hybrid approaches combine classical image processing methods with quantum computation. For instance, quantum circuits can help optimize specific enhancement tasks, such as contrast adjustment or color correction, allowing for more refined results.

To summarize, the future of image quality assessment is bright, with numerous potential breakthroughs and innovations on the horizon. By leveraging advancements in deep learning, real-time processing, multi-modal analysis, personalized assessments and quantum computing, IQA is poised to become more accurate, efficient and applicable across a wide range of industries. These innovations will not only enhance the quality of images but also improve user experiences and enable new applications in emerging technologies.

# References

[1]  M. Wang, "Blind image quality assessment: A brief survey," 2023. Available: https://arxiv.org/abs/2312.16551

[2]  W. Zhang, K. Ma, J. Yan, D. Deng, and Z. Wang, "Blind image quality assessment using a deep bilinear convolutional neural network," *IEEE Transactions on Circuits and Systems for Video Technology*, 30(1), 36–47, 2020.

[3]  K. Ma and Y. Fang, "Image quality assessment in the modern age," 2021. Available: https://arxiv.org/abs/2110.09699

[4]  H. Zhou, L. Tang, R. Yang, G. Qin, Y. Zhang, R. Hu, and X. Li, "Uniqa: Unified vision-language pre-training for image quality and aesthetic assessment," 2024. Available: https://arxiv.org/abs/2406.01069

[5]  Z. Ni, Y. Liu, K. Ding, W. Yang, H. Wang, and S. Wang, "Opinion-unaware blind image quality assessment using multi-scale deep feature statistics," 2024. Available: https://arxiv.org/abs/2405.18790

[6]  Z. You, J. Gu, Z. Li, X. Cai, K. Zhu, C. Dong, and T. Xue, "Descriptive image quality assessment in the wild," 2024. Available: https://arxiv.org/abs/2405.18842

[7]  S. Raviv and G. Chechik, "Assessing image quality using a simple generative representation," 2024. Available: https://arxiv.org/abs/2404.18178

[8]  C. Ozer, A. Guler, A. T. Cansever, and I. Oksuz, "Explainable image quality assessment for medical imaging," 2023. Available: https://arxiv.org/abs/2303.14479

[9]  Y. Chen, L. Liu, and C. Ding, "X-iqe: explainable image quality evaluation for text-to-image generation with visual large language models," 2023. Available: https://arxiv.org/abs/2305.10843

[10]  H. Zhao, H. Chen, F. Yang, N. Liu, H. Deng, H. Cai, S. Wang, D. Yin, and M. Du, "Explainability for large language models: A survey," 2023. Available: https://arxiv.org/abs/2309.01029

[11]  P. Cao, D. Li, and K. Ma, "Image quality assessment: Integrating model-centric and data-centric approaches," 2023. Available: https://arxiv.org/abs/2207.14769

[12]   Z. Yu, F. Guan, Y. Lu, X. Li, and Z. Chen, "SF-IQA: Quality and similarity integration for AI generated image quality assessment," in *Proceedings of the IEEE/CVF Conference on Computer Vision and Pattern Recognition (CVPR) Workshops*, June 2024, pp. 6692–6701.

[13]   S. Gu, J. Bao, D. Chen, and F. Wen, "Giqa: Generated image quality assessment," in *European Conference on Computer Vision (ECCV)*. Springer, 2020, pp. 369–385.

[14]   Y. Tian, Z. Ni, B. Chen, S. Wang, H. Wang, and S. Kwong, "Generalized visual quality assessment of GAN-generated face images," *arXiv preprint arXiv:2201.11975*, 2022.

[15]   Z. Yu, Y. Wang, Y. Zhang, L. Zhang, and X. Wang, "SF-IQA: Quality and similarity integration for AI generated image quality assessment," in *Proceedings of the IEEE/CVF Conference on Computer Vision and Pattern Recognition (CVPR) Workshops*, 2024.

[16]   X. Liu *et al.*, "NTIRE 2024 quality assessment of AI-generated content challenge," pp. 6337–6362, 2024, accessed on July 25, 2024. Available: https://openaccess.thecvf.com/content/CVPR2024W/NT IRE/html/Liu_NTIRE_2024_Quality_Assessment_of_AI-Generated_Conte nt_Challenge_CVPRW_2024_paper.html

[17]   A. Li, J. Wu, Y. Liu, and L. Li, "Bridging the synthetic-to-authentic gap: Distortion-guided unsupervised domain adaptation for blind image quality assessment," in *Proceedings of the IEEE/CVF Conference on Computer Vision and Pattern Recognition (CVPR)*, 2024, pp. 28422–28431.

[18]   L. Yang, H. Duan, L. Teng, Y. Zhu, X. Liu, M. Hu, X. Min, G. Zhai, and P. L. Callet, "Aigcoiqa2024: Perceptual quality assessment of AI generated omnidirectional images," *arXiv preprint arXiv:2404.01024*, 2024. Available: https://arxiv.org/abs/2404.01024

[19]   J. Wang, H. Duan, J. Liu, S. Chen, X. Min, and G. Zhai, "Aigciqa2023: A large-scale image quality assessment database for AI generated images: from the perspectives of quality, authenticity and correspondence," 2023.

[20]   P. Joshi, A. Rastegarpanah, and R. Stolkin, "A training free technique for 3D object recognition using the concept of vibration, energy and frequency," *Computers & Graphics*, 95, 92–105, 2021.

[21]   Frank Arute, Kunal Arya, Ryan Babbush, Dave Bacon, Joseph C. Bardin, Rami Barends, *et al.*, "Quantum supremacy using a programmable superconducting processor," *Nature*, 574, 505–510, 2019.

[22]   J. Gambetta, "The hardware and software for the era of quantum utility is here," *IBM Quantum research blog*, 2023.

[23]   Chao-Yang Lu, Zhang Jiang, Eric R. Anschuetz, Yunfeng Jiang, Hao Rong, *et al.*, "Evidence for the utility of quantum computing before fault tolerance," *Nature*, 618, 500–505, 2023.

[24]   S. Bravyi, O. Dial, J. M. Gambetta, D. Gil, and Z. Nazario, "The future of quantum computing with superconducting qubits," *Journal of Applied*

Physics, 132(16), 160902, 10 2022. Available: https://doi.org/10.1063/5.0 082975

[25]    A. T. e. a. Gupta R.S., Sundaresan N., "Encoding a magic state with beyond break-even fidelity," *Nature*, 625, 259–263, 2024.

[26]    Q. Liu, "Comparisons of conventional computing and quantum computing approaches," *Highlights in Science, Engineering and Technology*, 38, 502–507, 03 2023.

[27]    J. Biamonte, P. Wittek, N. Pancotti *et al.*, "Quantum machine learning," *Nature*, 549, 195–202, 2017.

[28]    T. Forcer, T. Hey, D. Ross, and P. Smith, "Superposition, entanglement and quantum computation," *Quantum Information & Computation*, 2, 03 2002.

[29]    Y. Wang, "When Quantum Computation Meets Data Science: Making Data Science Quantum," *Harvard Data Science Review*, 4(1), 2022, https://hdsr.mitpress.mit.edu/pub/kpn45eyx.

[30]    G. Ortolano, C. Napoli, C. Harney, S. Pirandola, G. Leonetti, P. Boucher, E. Losero, M. Genovese, and I. Ruo-Berchera, "Quantum-enhanced pattern recognition," *Physical Review Applied*, 20, 024072, 2023. Available: https://link.aps.org/doi/10.1103/PhysRevApplied.20.024072

[31]    A. Agrawal, M. Garg, S. Prakash, P. Joshi, and A. M. Srivastava, "Hand down, face up: Innovative mobile attendance system using face recognition deep learning," pp. 363–375, 2020.

[32]    P. Joshi and S. Prakash, "Image quality assessment based on noise detection," *2014 International Conference on Signal Processing and Integrated Networks (SPIN)*, pp. 755–759, 2014.

[33]    A. K. Singh, P. Joshi, and G. C. Nandi, "Development of a fuzzy expert system based liveliness detection scheme for biometric authentication," 2016.

[34]    S. K. *et al.*, "Pineapple sweetness classification using deep learning based on pineapple images," *Journal of Image and Graphics*, 11(1), 47–52, 2023.

[35]    J. Thomkaew and S. Intakosum, "Plant species classification using leaf edge feature combination with morphological transformations and sift key point," *Journal of Image and Graphics*, 11(1), 91–97, Mar 2023.

[36]    Y. Li, M. Tian, G. Liu, C. Peng, and L. Jiao, "Quantum optimization and quantum learning: A survey," *IEEE Access*, 8, 23 568–23 593, 2020.

[37]    A. Abbas *et al.*, "Quantum Optimization: Potential, Challenges, and the Path Forward," 12 2023.

[38]    H.-Y. Huang, M. Broughton, M. Mohseni *et al.*, "Power of data in quantum machine learning," *Nature Communications*, 12, 2631, 2021.

[39]    K. Rengasamy, P. Joshi, and V. Raveendra, "Hybrid facial expression analysis model using quantum distance-based classifier and classical support vector machine," 1, 1–6, 2023.

[40]    A. Callison and N. Chancellor, "Hybrid quantum-classical algorithms in the noisy intermediate-scale quantum era and beyond," *Physical Review A*, 106, 010101, 2022. Available: https://link.aps.org/doi/10.1103/PhysRevA. 106.010101

[41]   K. Rengasamy, P. Joshi, and V. Raveendra, "Real-time hybrid facial expression analysis model using quantum distance-based classifier and classical artificial neural networks: Quadcan," January 2024, pREPRINT (Version 1) available at Research Square.

[42]   M. Vallero, E. Dri, E. Giusto, B. Montrucchio, and P. Rech, "Understanding logical-shift error propagation in quanvolutional neural networks," *IEEE Transactions on Quantum Engineering*, pp. 1–15, 2024.

[43]   R. L'Abbate, A. D'Onofrio, S. Stein, S. Y.-C. Chen, A. Li, P.-Y. Chen, J. Chen, and Y. Mao, "A quantum-classical collaborative training architecture based on quantum state fidelity," *IEEE Transactions on Quantum Engineering*, pp. 1–13, 2024.

[44]   L. Le and T. N. Nguyen, "DQRA: Deep quantum routing agent for entanglement routing in quantum networks," *IEEE Transactions on Quantum Engineering*, 3, 1–12, 2022.

[45]   F. Tacchino, S. Mangini, P. Kl. Barkoutsos, C. Macchiavello, D. Gerace, I. Tavernelli, and D. Bajoni, "Variational learning for quantum artificial neural networks," *IEEE Transactions on Quantum Engineering*, 2, 1–10, 2021.

[46]   D. e. a. Domingo, Luis, "Binding affinity predictions with hybrid quantum-classical convolutional neural networks," *Nature Scientific Reports*, 13, 17951, 2023.

[47]   A. I. Gircha, A. S. Boev, K. Avchaciov, P. O. Fedichev, and A. K. Fedorov, "Hybrid quantum-classical machine learning for generative chemistry and drug design," *nature Scientific Reports*, 13, 8250, 2023.

[48]   D. Arthur and P. Date, "Hybrid quantum-classical neural networks," in *2022 IEEE International Conference on Quantum Computing and Engineering (QCE)*, 2022, pp. 49–55.

[49]   M. Schuld and N. Killoran, "Quantum machine learning in feature Hilbert spaces," *Physical Review Letters*, 122, 03 2018.

[50]   V. Havlíček,, A. D. Córcoles, K. Temme, A. W. Harrow, A. Kandala, J. M. Chow, and J. M. Gambetta, "Supervised learning with quantum-enhanced feature spaces," *Nature*, 567, 209–212, 2019.

[51]   Y. Lecun, L. Bottou, Y. Bengio, and P. Haffner, "Gradient-based learning applied to document recognition," *Proceedings of the IEEE*, 86(11), 2278–2324, 1998.

[52]   M. Henderson, S. Shakya, S. Pradhan, and T. Cook, "Quanvolutional neural networks: powering image recognition with quantum circuits," *Quantum Machine Intelligence*, 2(2), 2, 2020.

[53]   K. Rengasamy, P. Joshi, and V. Raveendra, "Real-time hybrid facial expression analysis model using quantum distance-based classifier and classical artificial neural networks: Quadcann," *Quantum Information Processing*, 23(6), 1–24, 2024.

# Index